# Cosmology Connections

## David Michalets

Self-published on **September 17, 2020**

# Table of Contents

Introduction . . . . . . . . . . . . . . . . . . . . . 4
1 Observing Our Universe Book . . . . . 8
2 Cosmology Transition Book . . . . . . . 11
3 Plasma . . . . . . . . . . . . . . . . . . . . . . . 19
4 Spectrum . . . . . . . . . . . . . . . . . . . . 21
5 Stars. . . . . . . . . . . . . . . . . . . . . . . . 35
6 Supernova. . . . . . . . . . . . . . . . . . . . 47
7 Plasmoid. . . . . . . . . . . . . . . . . . . . . 89
8 Neutron Star. . . . . . . . . . . . . . . . . . 105
9 Emission Nebula. . . . . . . . . . . . . . . 107
10 Nursery Nebula. . . . . . . . . . . . . . . 108
11 Planetary Nebula. . . . . . . . . . . . . . 112
12 Reflection Nebula. . . . . . . . . . . . . 130
13 Quasar. . . . . . . . . . . . . . . . . . . . . 135
14 Black Hole . . . . . . . . . . . . . . . . . . 138
15 Galaxy. . . . . . . . . . . . . . . . . . . . . . 144
16 Inter-Galactic Behaviors. . . . . . . . . 180
17 Solar System . . . . . . . . . . . . . . . . . 226
18 Final Conclusion . . . . . . . . . . . . . . 286
19 References . . . . . . . . . . . . . . . . . . 288

# Introduction

Cosmology has missed electrical connections.
The source of electromagnetic radiation is often incorrectly identified. This book offers a simple rule about synchrotron radiation to fix wrongly assigned sources.

Similarly, observations of distant objects result in explanations inconsistent with more probable causes. Electromagnetic forces and plasma behaviors can be ignored by a cosmology incorrectly expecting gravity and temperature can explain everything.

Our current, popular cosmology has several wrong assumptions causing some measurements to be used incorrectly, leading any progress by cosmology down a wrong path, resulting in more mistakes.
The author's first book, Observing Our Universe, detailed those fundamental mistakes.

The author's second book, Cosmology Transition, suggested many necessary changes to recover from those mistakes.
There were also suggestions for a new path for cosmology.

Among them are better models of a star, a spiral galaxy, and a quasar.

This third book offers more details about this progression for cosmology. The electrical and magnetic connections in the universe must be recognized, both in our solar system and beyond.

This book is the logical follow-up to the author's first two books by providing more observations and suggestions which cosmology must recognize. Otherwise, the science borders on irrelevance until it actually corrects its known mistakes.
This third book is clearer after the reader read the first two books to know both their conclusions and their justifications.

If those two books are not known to the reader, then some assumptions in this book may seem unjustified. Summaries are provided, but the books are not duplicated.

This book begins with a summary of the first two books in this series.

Note: the first book was completed before the second began. The second book was completed before this third book began. They are individual efforts with each having its own theme. One book never references a later book and one does not repeat material from an earlier book, or very little, when necessary.

Together the three books describe a transition to a "new and improved" comprehensive cosmology. Observations by the author are supplemented by theories developed by others. This is a brief summary of the respective sections:

1) Reviews the book: Observing Our Universe,

2) Reviews the book: Cosmology Transition,

3) Offers important details about plasma and electromagnetic forces,

4) Offers important details about a spectrum and electromagnetic radiation,

5) More suggestions about stars, beyond the second book.

6) Observations of a supernova,

7) Observations of a plasmoid,

8) Observations of a neutron star,

9) Observations of an emission nebula,

10) Observations of a nursery nebula,

11) Observations of a planetary nebula,

12) Observations of reflection nebulae,

13) Observations of a quasar,

14) Observations of a black hole,

15) More suggestions about galaxies.

16) Observations of Inter-galactic behaviors.

17) Observations in our solar system, including all of the planets.

18) Final conclusion for this book.

19) All references in this book can be found in a page in the author's web site, identified here.

Each page having an external reference has a link to the reference, whether a web page, pdf, or YouTube video.

A reader of the book with access to the Internet can read original reference material and view their high resolution images.

# 1 Observing Our Universe Book

The author's book, Observing Our Universe, was self-published using Kindle Book Publishing, so it is available from Amazon as either a 6x9 paperback or a Kindle download.

The book's description on Amazon:

We are Observing Our Universe with only one view, from on or near the Earth.
We are observing a spectrum with that particular line of sight to very distant objects.

We have the misperception of everything moving away, by a measured red shift.

That illusion can be explained.

There is a known crisis in cosmology, caused by the uncertain Hubble's Constant.

That crisis arose by not considering how our observations depend on our line of sight.
The Intergalactic Medium in our line of sight is important when observing distant galaxies.

The basics for a spectrum analysis and the Doppler Effect are thoroughly explained for galaxies, quasars, and stars.

We observe a gravitational wave indirectly, only by its effect on the Earth's surface.

Einstein's Theory of Special Relativity defined a special observer as one who is moving through a gravitational field, but we are not that special observer when here on Earth.

The context for every observation is important.

We cannot observe a black hole, or dark matter and dark energy.

We could not observe the big bang or its sequence, which were in the past.

Some claims are being made by cosmologists but with no evidence available to the public.

A theory requires a prediction and a search for evidence. Any evidence which conflicts with the prediction should force the theory's revision.

Evidence which confirms a prediction allows the theory to persist until conflicting evidence is observed.

Consistent evidence to the public is required for any claim to remain acceptable.

Evidence is lacking for some claims in cosmology and those crucial cases are explained, including their correct solution.

A red shift is not the only topic in the book, but it is the most important topic in cosmology, perhaps part of its foundation.

The correct understanding of a red shift is crucial.

The notorious Hubble's Constant which is based on observing red shifts is at the root of the current crisis in cosmology.

The conclusions of this book impact astrophysics to some extent.

The topics collectively reveal the prominent, though not publicized, errors in modern cosmology and how to fix them. (End description)

Book's main topics were Red Shifts, Gravitational Waves, and Relativity.


The book's simplified conclusion:

After cognizing redshifts of galaxies and quasars are never their true velocity, nearly everything in cosmology must change. All the galaxies distances must be adjusted for relevant Cepheid variables to correctly account for the inter-galactic medium. Quasars have no method to measure their distance.

Relativity cannot apply to the context of the universe, despite claims to the contrary.

The author created a separate pdf of the problem with relativity in cosmology. Its title is Removing Relativity, based on the first book's content.

A link to that document, Removing Relativity, is available in the section for References at the end oft this book.

# 2 Cosmology Transition Book

The author's book, Cosmology Transition, was self-published using Kindle Book Publishing, so it is available from Amazon as either a 6x9 paperback or a Kindle download.

The book's description on Amazon:

Cosmology is in Transition. This book follows the book Observing Our Universe, which marked the need for a transition in cosmology.
The first book explained several crucial mistakes in cosmology, some dating back 100 years, including the one which resulted in dark energy,.
This second book explains how cosmology can transition after fixing the mistakes and their consequences, and then considering alternatives for a better cosmology.
The first book mentioned possible changes. This book describes some of the available alternatives for this transition in cosmology.

The book weaves several threads to propose a new path for a new cosmology.
Among those threads are:
a) A new solar model proposed by Dr. Robitaille, but this model is based on condensed matter, not gaseous plasma,
b) a model for a spiral galaxy based on one proposed by Hannes Alfven,
c) The author's quasar model,

d) The replacement of relativity with the appropriate mix of the forces of electric, magnetic, and gravity. Gravity alone fails in cosmology to explain all motion, beyond the solar system.
Relativity enabled the false concept of a black hole, which arose only for the moving observer but then cosmology claimed this space-time could be applied to the universe. Relativity fails in cosmology. The claimed M87 black hole was really a plasmoid.

Hannes Alfven was awarded the 1970 Nobel Prize in Physics for his important work developing plasma physics. Cosmology made a mistake when not adopting this crucial advance in physics. Nearly all the universe is plasma, or matter having an electrical charge, but cosmology maintained its reliance on the weak force of gravity. The result was the dark matter mistake.

This book offers alternatives for this transition. This is not a definitive plan. There is more than one alternative and this author cannot claim with certainty a correct path, but practical alternatives are offered. Our solar system is understood but there are problems beyond our Local Group. This transition to recover from fundamental mistakes has several clear alternatives for this time of transition.

Perhaps this transition will be difficult.
After the double debacle of dark matter and dark energy, cosmology needs suggestions.
(End description)

Book's main topics were updating astronomical data, new solar model, new spiral galaxy model, new quasar model.

The book's simplified conclusion:

After recognizing redshifts of galaxies and quasars are never their true velocity, nearly everything in cosmology, based on that wrong data, must change. All the galaxy distances must be adjusted for relevant Cepheid variables to correctly account for the inter-galactic medium. Quasars have no method to measure their distance or velocity.

There is no expansion and no big bang.

Relativity cannot apply to the context of the universe, despite claims to the contrary.

Reference section has a link to Removing Relativity pdf for a summary of that topic in the book.

# Red Shifts with a Simple Explanation

The term "red shift" is used so loosely, most think of it as just a simple number having a consistent meaning, like a temperature.

A red shift is not that simple and anyone using the term so loosely is showing they consider it as just a simple number.

It is crucial to recognize there are 4 different red shifts. Each is a measurement of a distinct behavior.

Galaxies are totally different entities than quasars. A galaxy has billions of stars while a quasar is a quasi-stellar object having no stars.

A metallic element is one which is not hydrogen or helium.

The 4 distinct red shifts and mechanisms:

1) galaxy – hydrogen

2) galaxy – metal

3) quasar – hydrogen

4) quasar – metal

Red shift (1) the hydrogen absorption line is driven by hydrogen in the inter-galactic medium. This line is not from the galaxy.

Red shift (2) the calcium ion absorption line is driven by calcium ions near the galactic corona, as in the case of M31 and others. Calcium is a metal. The metallic line is not from the galaxy.

Red shift (3) the quasar high red shift comes from the hydrogen Lyman-alpha emission line.

Red shift (4) the quasar low red shift comes from the metallic ion emission lines.

All red shifts in the list can never be a velocity of the object. This is crucial.

That mistake resulted in the false expansion.

However, when Red shift (1) is used in conjunction with a Cepheid in the galaxy, this value enables a rough distance calculation. Using a Cepheid for a distance value is a distance metric becomes available for the hydrogen density within the IGM in the line of sight to its galaxy. The highest red shift (1) has been measured at $z > 11$, so clearly the Red shift (1) z value cannot represent a velocity.

For red shift (2) there are galaxies with either a red or blue shift of the metallic ion absorption lines. This value cannot be used for a distance calculation.

LINER galaxies, which include Seyferts, exhibit several metallic elements when taking the spectrum of only the AGN. None of these metallic lines in a LINER galaxy spectrum are related to the galaxy distance.

For red shift (3) this hydrogen emission line is found in a "typical" quasar. This line can never indicate a quasar velocity, nor can it be related to a quasar distance.

For red shift (4) these metallic lines are found in the quasars used by Halton Arp, in his book Seeing Red. This can never be a quasar velocity, nor can it be related to a quasar distance, nor can it be related to the age of matter. These ions slow down in apparent incremental changes, in their velocity, as the cloud disperses.

The z value for red shift (3) has exceeded 7, while the z value for red shift (4) is usually < 1. The values are never related to distance.

It is crucial to note that none of the 4 types of a red shift is an indicator of the object's real velocity.

When one accepts that simple fact about the false velocities, then there is no "Hubble Flow." That was the term Edwin Hubble used initially for the red shift trend, but later in 1936, he noticed red shift (1) is observed with only galaxies beyond our Local Group.

Hubble recognized the "Hubble Flow" was not consistent. Dark energy arose from the wrong assumption that the false expansion is consistent.

There is also no expansion, no dark energy, and no big bang. These are not topics for this book. Those were covered in the earlier books.


Any concept in cosmology which treats a red shift as just a number, ignoring there are actually 4 types of red shifts, is immediately wrong. Cosmological red shift is one example of that mistake. When assuming all red shifts are the same mechanism.

Another important observation is X-rays are at the high end of synchrotron radiation.

Thermal radiation spans from infrared to ultraviolet.

X-rays are never from thermal radiation because such an extreme temperature is impossible.

Similarly, radio is at the very low end of synchrotron radiation, so that is its only source.

Reference section has a link to Clarifying Redshifts pdf which was the original basis for that section of the first book.

As a conclusion of the first two books, there are no galaxies and quasars with known 3-dimensional velocities. Cosmology must cease its double false assumption they are known and all are directly away from Earth. As a result, nearly all published distances are wrong.

Among the crucial recommendations for the cosmology transition is all the astronomical data about velocities and distances must be updated for the correct usage of the 4 types of red shifts.

For the transition in astronomical data, the author offers a data base design as an initial suggestion. Astronomers must contribute their requests in its development. Currently numbers are published with no method for their verification or correction with subsequent observations.

# 3 Plasma

Excerpt from plasma universe site:

A plasma (often ionized gas), is a gaseous substance consisting of free charged particles, such as electrons, protons and other ions, that respond very strongly to electromagnetic fields. The free charges make the plasma highly electrically conductive, so that it may carry electric currents, and generate magnetic fields. This may cause the plasma to constrict (or pinch) into filaments, generate particle beams, emit a wide range of radiation (radio waves, light, microwave, x-ray, gamma and synchrotron radiation), and form cellular regions of plasma with similar characteristics (e.g. magnetosphere, interplanetary medium).

Plasma is usually considered to be a distinct phase of matter from solids, liquids, and gases because of its unique properties, that it is often called the "fourth state of matter", or even the "first state of matter"

Plasma typically takes the form of neutral gas-like clouds or charged ion beams, but may also include dust and grains, called dusty plasmas. They are typically formed by heating and ionizing a gas, stripping electrons away from atoms, thereby enabling the positive and negative charges to move freely.

A solar coronal mass ejection blasts plasma throughout the solar system.

Plasma was first identified in a discharge tube (or Crookes tube), and so described by Sir William Crookes in 1879 (he called it "radiant matter") The nature of the Crookes tube "cathode ray" matter was subsequently identified by British physicist Sir J.J. Thomson in 1897, and dubbed "plasma" by Irving Langmuir in 1928, perhaps because it reminded him of a blood plasma. Langmuir wrote:

"Except near the electrodes, where there
are sheaths containing very few electrons, the ionized gas contains ions and electrons in about equal numbers so that the resultant space charge is very small. We shall use the name plasma to describe this region containing balanced charges of ions and electrons."

(Excerpt end)

# 4 Spectrum

Spectrum analysis is the basis for understanding all objects beyond our solar system.

The word light will be used for electromagnetic radiation, which is the propagation of perpendicular, synchronized, electric and magnetic fields. These fields oscillate at a rate defined either as a frequency (measured as cycles per second or hertz) or as a wave length (usually measured in nanometers, or $10^{-9}$ m, or Angstroms, or $10^{-10}$ m). The propagation of these oscillating fields has been measured in a vacuum using our standard for units of time and is called the constant c. The velocity is reduced in a medium, as defined by the medium's diffraction index.

When the path of an electron or an electric current is diverted by a magnetic field, the result is synchrotron radiation because that is how a synchrotron works. Synchrotron radiation covers a broad spectrum of frequencies. The range depends on the velocity of the electron or the intensity of the electric current. A simple behavior of electrodynamics is a collapsing electric field creates a magnetic field, and a collapsing magnetic field creates an electric field. The propagation continues until the radiation is absorbed.

The highest frequency or shortest wave length in this flat distribution is determined by the electron velocity. Fast electrons or an electric current can achieve X-rays or even gamma rays for the highest energy radiation.

It must be noted gamma rays are observed from extreme lightning bolts here on Earth.

Excerpt from Wikipedia about synchrotron radiation:

When high-energy particles are in acceleration, including electrons forced to travel in a curved path by a magnetic field, synchrotron radiation is produced.

It is considered to be one of the most powerful tools in the study of extra-solar magnetic fields wherever relativistic charged particles are present. Most known cosmic radio sources emit synchrotron radiation. It is often used to estimate the strength of large cosmic magnetic fields as well as analyze the contents of the interstellar and intergalactic media.

(Excerpt end)

Thermal radiation is characterized by a frequency distribution where the frequency with the highest intensity is related to the object's temperature. Our Sun's thermal spectrum is shown in a figure below.

Thermal radiation extends from ultraviolet to infrared frequencies. The visible range of frequencies is roughly in the middle of this range.

When an object is emitting thermal radiation it will be felt as heat if hot enough because our sense of touch feels the infrared frequency as heat. When its temperature is hot enough for the frequency of the color red to be in the spectrum, then we can see the object as red.
If its temperature gets even hotter, then the mix of colors in the spectrum will be seen as white. When hot enough to be mostly in the ultraviolet, that frequency range is not visible.

Our Sun changes color from yellow overhead to reddish near the horizon due to the density of the atmosphere in the line of sight. When beyond the atmosphere our Sun appears white.

The first book had an extensive description of a spectrum, absorption and emission lines.

Most emission lines from the elements are in the ultraviolet range but a few are visible.

If an object's spectrogram has wave lengths beyond ultraviolet (or X-ray and gamma ray) or below infrared (or radio) then that part of the spectrum cannot be thermal radiation.

When a spectrum has non-thermal radiation then it must be synchrotron radiation, which is the result of moving electric charges diverted by a magnetic field.

Radio emissions from the planet Jupiter were first detected in 1955. Jupiter is known to have a strong magnetic field and the planet is a source of low energy synchrotron radiation.

There are several light sources:

a) A star is the ubiquitous light source in the universe,

b) a quasar is a quasi-stellar object, meaning it looks like a very distant star but it generates synchrotron radiation extending from X-ray to radio frequencies. Quasars are often shrouded in clouds of atoms. The visible range of frequencies is often dimmed from a quasar.

There are several celestial entities containing many stars:

a) Globular cluster, which is essentially a small sphere of stars with a rough diameter.

The spheres of stars do not rotate.
c) Elliptical galaxy, which is a huge sphere of stars, sometimes over a trillion stars.

This sphere can be somewhat flattened from a perfect sphere to look like an ellipse. All the different shapes are called an elliptical galaxy,

d) Spiral galaxy, which has a central bulge of stars amidst a rotating disk of stars, dust, and clouds of atoms and molecules. The disk can have one or more arms which appear to spiral from the central bulge during the rotation of the disk.

e) Lenticular galaxy which has a shape between a spiral and an elliptical because it has the central spherical bulge of stars and a disk of dust and stars but having no arms and no rotation.

f) Ring galaxy which is the rare combination of the central bulge of stars surrounded by a ring of stars and dust with a gap between bulge and ring.

g) Irregular galaxy is one which does not have a shape like any of the defined shapes above.

Globular clusters are associated with any of the galaxy types. Larger galaxies will have a higher number of globular clusters.

h) BL Lac object, named for its first instance, BL Lacertae, a star in the constellation Lacerta. It is a source of synchrotron radiation like a quasar but has no clouds of atoms, so its core has no dimming. This object can also be called a plasmoid (in its own section), a term coined in the 1950's by Winston Bostwick, a plasma physicist.

i) Atom or molecule.

An atom or ion will emit light at a specific wavelength defined by the atom's electron configuration when it captures an electron.

Thermodynamics describes the process how energy can be transferred or changed between bodies or changed to another form.

Various frequencies and wave lengths will be mention.

Here is a reference for the numbers and units.
A nm or nanometer is $10^{-9}$ m

An Angstrom is 0.1 nanometer or $10^{-10}$ m

A spectrogram, or the intensity of each wavelength in the measured spectrum, typically has Angstroms for the units on the horizontal axis, and a normalized intensity value on the vertical axis.

First, Hydrogen's details:

The Lyman series is associated with the n=2 orbital in Hydrogen while the Balmer series is for the n=3 orbital.

Lyman-alpha, Ly-α, Lyman-α, K-alpha, K-α are all 1215.67 Angstroms.

Hydrogen-alpha, Ba-α, Balmer-alpha are all 6563 Angstroms.

Lyman-beta, Ly-β are 1025.722 Angstroms.

Hydrogen-beta, Ba-β are 4861 Angstroms.

Rarely encountered:

Lyman-gamma, Ly-γ, are 972.537 Angstroms

Hydrogen-gamma, Hydrogen-γ, Balmer-gamma are 4340 Angstroms.

A crucial possible mistake is confusing Ly-α with Hydrogen-α because these "alphas" are different coming from a different orbital series.

Here are several named ranges and their wavelengths:

gamma rays 1pm or E-12 m
hard x-rays 10pm or E-11
soft x-rays 100 pm or E-10 or Angstroms
extreme UV 10 nm or E-9
near UV 100 nm or E-7
UVC 100-280 nm
UVB 280-315 nm
UVA 315-400 nm

visible colors 380-750 nm
visible colors in Angstroms:

Violet 4000
Blue 4600
Cyan 4900
Green 5000
Yellow 5800
Orange 6000
Red 7000

near infrared 1 um or E-6 m
mid infrared 10 um or E-5 m
far infrared 100 um or E-4 m
EHF 1 mm or E-3 m
SHF 10 mm or E-2 m
UHF 100 mm or E-1 m
VHF 1 m
HF 10 m
MF 100 m

LF 1km
VLF 10 km
ULF 100 km
SLF 1000 km or E+6 m
ELF E+7 m


NIST element lines are in Angstroms = 1E-10 m or 0.1 nm.

Selections are by NIST intensity at 1000

Argon 671.8513 A, 723.3606 A, 1048.1987 A

NIST table (can select an element with periodic Table button near top using the periodic table)

helium 303.7804, 584.33436 A

Hydrogen Lyman-alpha 1215.66824 A or 121.6 nm (UV)

Wikipedia uses nm:

hydrogen alpha 656.45337 nm or Ballmer-alpha - red
hydrogen beta 486.13615 nm or Ballmer-beta - aqua
hydrogen gamma 434.0462 nm or Ballmer-gamma - blue
hydrogen delta 410.174 nm or Ballmer-delta - violet
hydrogen epsilon 397.0072 nm or Ballmer-epsilon - UV
hydrogen zeta 688.9049 nm or Ballmer-zeta -UV
hydrogen nu 383.5384 nm or Ballmer-nu - UV

Lyman limit or Lyman Break 912 nm

Others:

Al II 1671
Al III 1863

Fe II 2382
Mg II 2798

Ne III 3869
Ne V 3426

O II 3727
O III 4959
O III 5007

C II 1335
C IV 1548.195, 1550.770

Fe II 2382.765, 2600.173
Mg II 2796.352, 2803.531

Si IV 1393.755, 1402.770
Si II 1527

calcium 3933.6614, 3968.4673 A

carbon 687.346 A, 1930.906 A

Iron 2382.0376, 2483.2708 A
magnesium 2795.5301, 2852.1251 A

neon 460.7284 A

nitrogen 645.179, 1199.550 A

oxygen 1302.168 A

potassium 495.14 A, 612.62 A
potassium 7664.8991, 7694.9645 A

silicon 1264.7379, 1988.9937, 2881.5771 A

sulfur 1425.030, 1473.995, 1807.311 A

tellurium 2002.028, 4686.91 A

tin 1290.880 A, 1400.440, 1474.997 A

titanium 3349.405 A

Zinc 2025.4845 A, 2062.011 , 2099.9273 , 2138.5735 A

For an academic interest in hydrogen, the Paschen series (with n from 4 to 3) begins with its alpha at 1875 nm or 18,750 Angstroms, or in infrared.
The hydrogen Brackett series ( n from 5 to 4) begins with its alpha at 4051 nm or 40,510 Angstroms, or in far infrared.

Values are from several sources, such as onc with a spectrogram (usually a quasar) having those specific element lines identified.

References at the end of this book has a link to the NIST table of elements, with calcium pre-selected.

Following are several examples of spectrograms.

1) our Sun,
2) M31 galaxy,
3) typical quasar.

From Wikipedia, here is the Sun's thermal radiation:

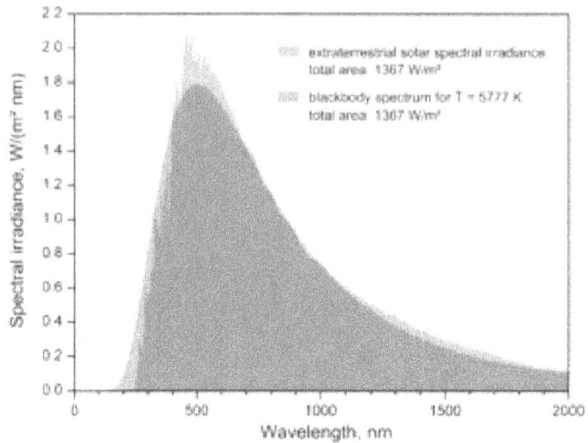

The caption:

The effective temperature, or black body temperature, of the Sun (5777 K) is the temperature a black body of the same size must have to yield the same total emissive

This M31 spectrogram is found by:
"The radial velocity measure of nearby galaxies"

The following image is from that page:

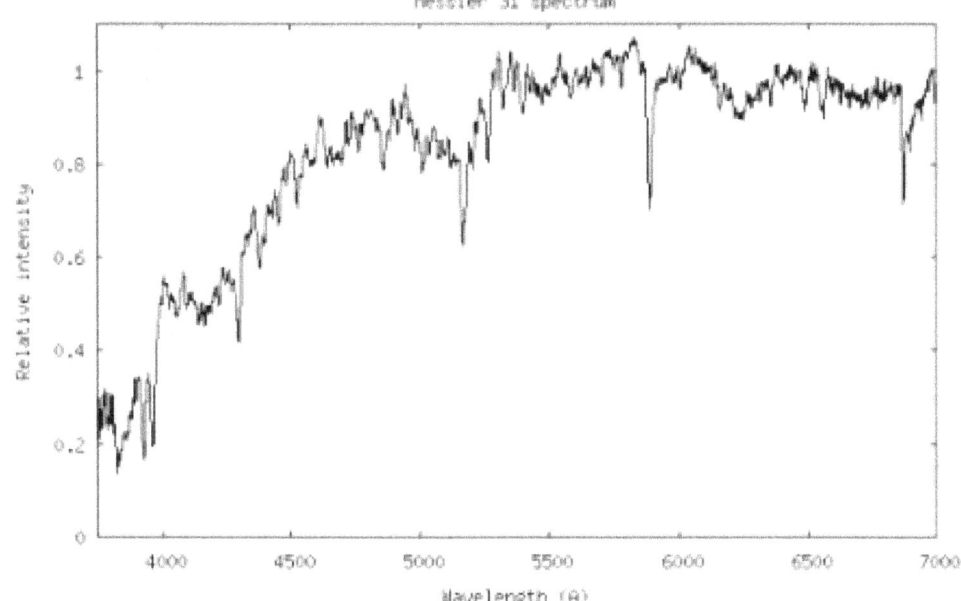

This is clearly the somewhat flat distribution of frequencies observed with synchrotron radiation. If this were from a star like our Sun, then its distribution cannot be flat like this.

The entire spectrum of synchrotron radiation from the galactic corona spans the entire spectrogram.

M31 has the absorption lines for calcium ions (below 4000 A). They show a Doppler Effect velocity of -301 km/s. This is about the same velocity of the calcium ions moving away from the solar corona.

The spectrum clearly shows the corona has a source of synchrotron radiation. The page also includes spectrograms for 3 stars (in blue, green, red) have the distinctive hump of thermal radiation for the hottest star. Each star slopes down to around 7000 Angstroms, or infrared.

A spiral galaxy has a bulge of stars around the central core. By this observation, there is an electrical current (moving charged particles) present near the corona. It is changing its path by a magnetic field resulting in the propagation of electromagnetic radiation. With this mechanism, it is called synchrotron radiation because that is how a synchrotron device works.

This spiral galaxy corona is crucial when measuring the galaxy spectrum. If only stars are in the sample, they generate thermal radiation. If the corona is in the sample, it flattens the frequency distribution as observed in the M31 spectrogram.

The calcium ions causing their absorption lines are in the line of sight to the M31 galactic corona. As noted in the first book, these absorption lines indicate nothing about a velocity of M31.

The spectrum of the corona slice is clearly not of a star. It is not from a gas, which can generate only emission lines from the individual ions.

The galactic corona of a spiral galaxy like M31 is a source of synchrotron radiation.

a Caltech study in 2000 provided a "typical spectrum for a quasar" with z=1.34.

This study can be found with
Quasistellar Objects: Intervening Absorption Lines"

Its figure 1 is important for a comparison to galaxy.

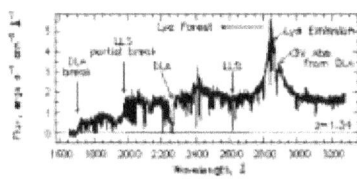

**Figure 1.** Typical spectrum of a quasar, showing the quasar continuum and emission lines, and the absorption lines produced by galaxies and intergalactic material that lie between the quasar and the observer. This spectrum of the $z \sim 1.34$ quasar PKS0454−039 was obtained with the Faint Object Spectrograph on the Hubble Space Telescope. The emission lines at ~ 2400 Å and ~ 2850 Å are Lyβ and Lyα. The Lyα forest, absorption produced by various intergalactic clouds, is apparent at wavelengths blueward of the Lyα emission line. The two strongest absorbers, due to galaxies, are a damped Lyα absorber at $z \sim 0.86$ and a Lyman limit system at $z \sim 1.15$. The former produces a Lyman limit break at ~ 1700 Å and the latter a partial Lyman limit break at ~ 1950 Å since the neutral Hydrogen column density is not large enough for it to absorb all ionizing photons. Many absorption lines are produced by the DLA at $z \sim 0.86$ (C IV λλ1548, for example, is redshifted onto the red wing of the quasar's Lyα emission line).

# 5 Stars

For those not reading the first 2 books: A star being composed of condensed matter, not a gas, is the apparent consensus in a new cosmology, as it is part of both Electric Cosmology and Plasma Cosmology.

## 5.1 Sun is composed of condensed matter

Dr. Pierre-Marie Robitaille began his publications of his condensed matter theory of the Sun with a paper in 2013, titled:

Forty Lines of Evidence for Condensed Matter - The Sun on Trial: Liquid Metallic Hydrogen as a Solar Building Block"

A star is composed of metallic hydrogen, a lattice of protons and electrons, which is a form of condensed matter. The Sun is not a sphere of gas as claimed by many over the last century, though a few proposed a liquid sun.

There is a video titled: Sun is not gaseous.

The photosphere is an actual liquid surface, not an illusion in the solar atmosphere as proposed in the gaseous sun model.

The core is solid metallic hydrogen which is heated by the axial solar electric current. This heat is moved by convection to the photosphere which cools by its thermal radiation.

Above the photosphere, sunspots, faculae, and other behaviors like the solar wind and coronal mass ejections are driven by magnetohydrodynamics with this mix of magnetic fields and liquid metallic hydrogen.

Transmutation, also known as cold fusion, occurs on the surface of a star. There is no fusion in its core. That fusion process requires extreme pressures and temperatures, and a precarious equilibrium to avoid gravitational collapse. Dr. Pierre-Marie Robitaille has done presentations at Electric Universe or Thunderbolts conventions.

## 5.2 Life and death of electric star

He also has a presentation at a Suspicious Observers convention in 2017.

This video is titled "Life and death of electric star"

## 5.3 The HR Diagram Explained

He also has a video titled "Different Star Types? The HR Diagram Explained!"

Figure from Wikipedia:

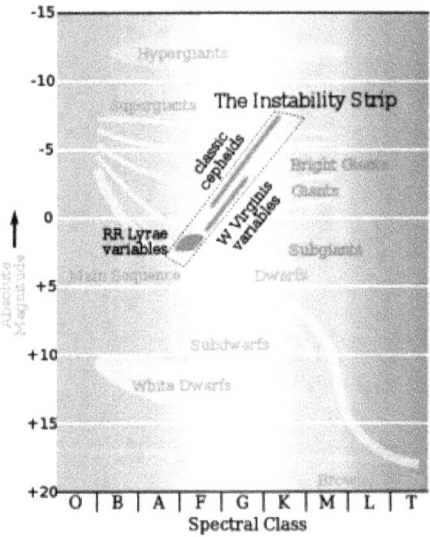

At the end of this book is a page to direct the reader to a web page offering links to online documents, articles, and videos.

## 5.4 Solar circuit

For those who don't have the second book, this text and figure were included to show the electrical connection to a star.

From Wikipedia:

Between 1994 and 1995 [Ulysses space craft] explored both the southern and northern polar regions of the Sun, respectively.

(Excerpt end)

The following figure is from the "holoscience" site and its page titled:
"Alfven Triumphs Again ( &Again)"

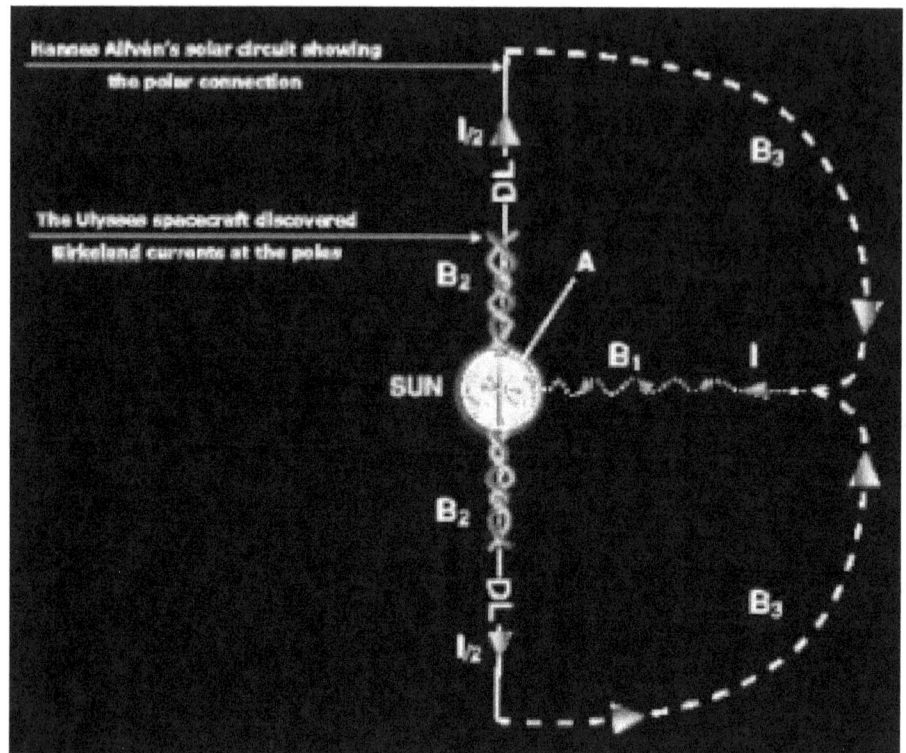

The caption:

Alfvén's Heliospheric Circuit. The Sun acts as a unipolar inductor (A) producing a current which goes outward along both the axes (B2) and inward in the equatorial plane along the magnetic field lines (B1).

The current must close at large distances (B3), either as a homogeneous current layer, or — more likely — as a pinched current. Analogous to the auroral circuit, there may be double layers (DLs) which should be located symmetrically on the Sun's axes. Such double layers have not yet been discovered. Credit: Original diagram by H. Alfvén, NASA Conference Publication 2469, 1986, p. 27.

(End caption)

One crucial observation is the pair of double layers at the North and South poles of a star.

If the DLs collapse, one possible result is the ejection of the star's photosphere, its outer layer. This is a possible scenario for the ejection of the plasma sphere called a planetary nebula.

The following figure comes from the YouTube video titled:

Beyond Juergens Electric Sun with Wal Thornhill & Don Scott: The Solar Circuit & Sunspots

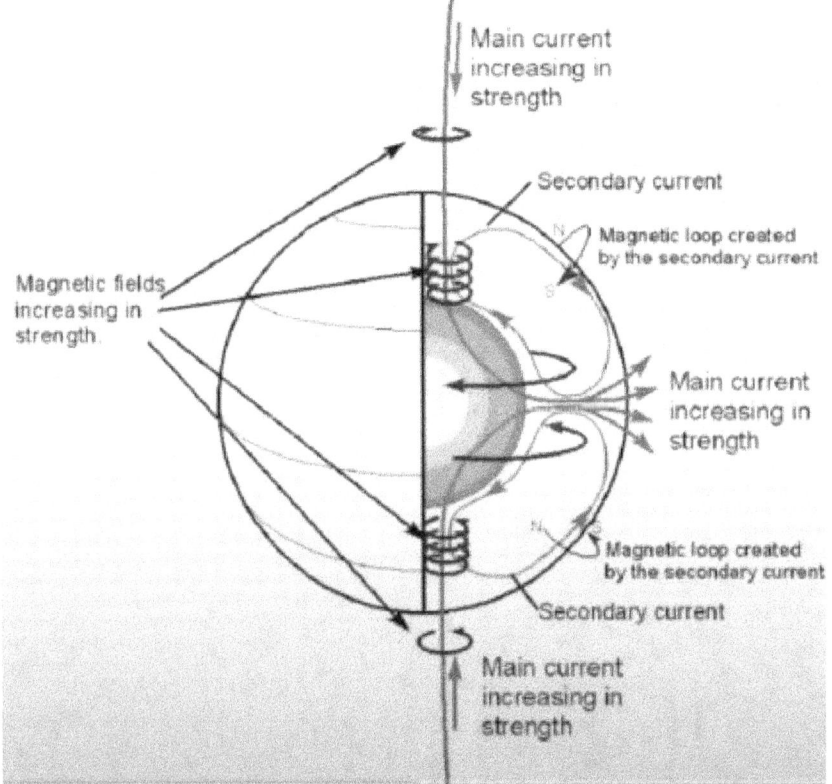

One critical observation is the electric current loops below the surface and in opposite directions in either hemisphere.

The solar sunspots are always opposite N-S polarity between hemispheres. These polarities flip each solar cycle. These behaviors are driven by sub-surface currents and their direction.

The axial Birkelund currents carry currents in opposing directions. This behavior is similar to plasma filaments which carry both positive and negative charges together.

The Sun has many separate electric currents while each generates a magnetic field. These behaviors are described in the video.

There is no transcript available for an excerpt.

## 5.5 Pulsar

Excerpt from Wikipedia:

A pulsar (from pulse and -ar as in quasar) is a highly magnetized rotating neutron star that emits beams of electromagnetic radiation out of its magnetic poles. This radiation can be observed only when a beam of emission is pointing toward Earth (much like the way a lighthouse can be seen only when the light is pointed in the direction of an observer), and is responsible for the pulsed appearance of emission. Neutron stars are very dense, and have short, regular rotational periods. This produces a very precise interval between pulses that ranges from milliseconds to seconds for an individual pulsar. Pulsars are one of the candidates for the source of ultra-high-energy cosmic rays (see also centrifugal mechanism of acceleration).
(Excerpt end)

Observation:
Both the density of a star consisting of only neutrons and the extreme rotation for this "lighthouse" effect are impossible.

Thunderbolts Project has an explanation for the object with pulsar behaviors which is not a neutron star.
The periodic variation in radiation is an electrical behavior, like a capacitor charging and discharging, not a mechanical behavior, like a lighthouse.

Space News has a pulsar video titled:

"Impossible" Pulsar Breaks the Rules | Space News


Chandra Solves a Pulsar Puzzle

This story is titled "Chandra Images Show That Geometry Solves a Pulsar Puzzle"

This story from Chandra is describing two similar pulsars, both "rotating about five times per second."

Excerpt from the story about two similar pulsars but in slightly different orientations:

A likely interpretation of the Chandra images is that the long narrow trails to the side of Geminga and the double tail of B0355+54 represent narrow jets emanating from the pulsar's spin poles. Both pulsars also contain a torus, a disk-shaped region of emission spreading from the pulsar's spin equator. These donut-shaped structures and jets are crushed and swept back as the pulsars fly through the Galaxy at supersonic speeds.

In the case of Geminga, the view of the torus is close to edge-on, while the jets point out to the sides. B0355+54 has a similar structure, but with the torus viewed nearly face-on and the jets pointing nearly directly towards and away from Earth. In B0355+54, the swept-back jets appear to lie almost on top of each other, giving a doubled tail.

Both pulsars have magnetic poles quite close to their spin poles, as is the case for the Earth's magnetic field. These magnetic poles are the site of pulsar radio emission so astronomers expect the radio beams to point in a similar direction as the jets. By contrast the gamma-ray emission is mainly produced along the spin equator and so aligns with the torus.

For Geminga, astronomers view the bright gamma-ray pulses along the edge of the torus, but the radio beams near the jets point off to the sides and remain unseen. For B0355+54, a jet points almost along our line of sight towards the pulsar. This means astronomers see the bright radio pulses, while the torus and its associated gamma-ray emission are directed in a perpendicular direction to our line of sight, missing the Earth.

(Excerpt end)

Observation:
The pair of images in the story helps with the descriptions (above and below).
The two images: top left < 2ly at distance of 3400ly, top right about 1ly at distance 800ly.

The top left pulsar clearly has jets out both poles.
The jet out to the above left splits in two and arcs out and down symmetrically but then both cease their X-ray emissions.
The jet out to the lower right is linear but its X-ray emissions are not consistent before dimming.

The top right pulsar has a bowl shape with the pulsar at the base of this bowl. X-ray emissions are also stronger along the rim of this bowl shape. There is also a wide jet of X-ray emissions symmetrical to this bowl, to the lower right, in line with the pulsar.

Both the half circle or the torus shape are due to the plasmoid's magnetic field, not that either is "crushed" by the velocity of the object.

These pulsar observations are consistent with the M87 plasmoid description.

# 6 Supernova

Excerpt from Wikipedia:
A supernova occurs during the last evolutionary stages of a massive star or when a white dwarf is triggered into runaway nuclear fusion. The original object, called the progenitor, either collapses to a neutron star or black hole, or is completely destroyed. The peak optical luminosity of a supernova can be comparable to that of an entire galaxy before fading over several weeks or months.

(Excerpt end)

Observation:

The supernova event involves the ejection of a star's outer layers but never finishes with the fictitious entities of black hole or neutron star.

## 6.1 Supernova Type IA Assumptions

Supernova type IA events are hoped to be used as standard candle. The observed light curve is assumed to allow a direct calculation of its distance. Unfortunately, these "super explosions" never have a consistent luminosity curve.

Quotes and data are from Wikipedia.

Supernovae are rare:

Only three naked-eye supernova events have been observed in the Milky Way during the last thousand years. The most recent directly observed supernova in the Milky Way was Kepler's Supernova in 1604.
Observations of supernovae in other galaxies suggest they occur in the Milky Way on average about three times every century.
(Excerpt end)

Observation:

The one before 1604 was in 1572.
With the last one over 4 centuries ago this expected rate of 1/33yr is inaccurate. None of the expected 12 were detected in this span.
Our galaxy offers a very minimal history.

The most recent supernova IA observed in the universe after 1604 was SN1937C in IC 4182 in the Canes Venatici constellation, but Wikipedia has few details.

The next IA after that one in 1937 was SN1972E in NGC 5253 in the Centaurus constellation, and "became the prototypical Type Ia supernova."

Now that astronomers are looking for these IA type supernovae, others have followed (Hubble launched in 1990):

SN1994D in NGC 4526 in the Virgo constellation,

SN2002bj in NGC 1821 in the Lupus constellation,

SN2003fg in "anonymous galaxy" in the Bootes constellation,

SN2004dj in NGC 2403 in the Camelopardalis constellation,

SN2009-MENeaC in the Aries constellation(this one was observed in a globular cluster near an anonymous elliptical galaxy in a cluster at a distance of 1 billion lyr),

SN2010lt in UGC 3378 in the Camelopardalis constellation.

Since 1972 defined the prototype for IA, only 6 have followed.

This is an interesting diversity including associations with anonymous galaxies, even one within a globular cluster.

Excerpt:

Theoretical studies indicate that most supernovae are triggered by one of two basic mechanisms: the sudden re-ignition of nuclear fusion in a degenerate star or the sudden gravitational collapse of a massive star's core.

In the first class of events, the object's temperature is raised enough to trigger runaway nuclear fusion, completely disrupting it. Possible causes are accumulation of sufficient material from a binary companion through accretion, or a merger. In the massive star case, the core of a massive star may undergo sudden collapse, releasing gravitational potential energy as a supernova. While some observed supernovae are more complex than these two simplified theories, the astrophysical mechanics have been established and accepted by most astronomers for some time.

(Excerpt end)

Observation:
The two simplified theories involve a disruption in the fusion cycle.
The massive star releasing 'gravitational potential energy' to become a supernova makes no sense.
Work can be done on a body lifting it a distance in a gravitational field.
The body is said to have gravitational potential energy in that state.
When the body is released it moves at free fall acceleration toward the body with this gravitational field.
after the body is released, that gravitational potential energy has no affect.

Excerpt:
gravitational potential energy is most useful for calculating the escape velocity from the earth's gravity.
(Excerpt end)

Observation:
In the suggested massive star scenario, part of the star has somehow separated, is being held apart, then released to accelerate in free fall toward the larger portion resulting in a violent collision, observed as a supernova.
This substantial internal separation cannot be possible in a massive star under such extreme density and pressure to sustain fusion.

No explicit mechanism in a supernova is suggested other than "runaway nuclear fusion."
Any use of "runaway" means this proposal violates known physics.

The 1998 paper about using supernovae to measure a false accelerating universe expansion was titled:

Dark Energy, and the Accelerating Universe: The Status of the Cosmological Parameters

Excerpt:

To understand how the studies deal with the questions of supernova evolution, it is important to be clear that evolution is not assumed to be a monotonic function of the age of the universe; in other words, we do not expect that supernovae are uniformly fainter as you study them at higher redshifts. Rather, the main concern is that the typical environment in which a supernova explodes may on average be a little bit different at high redshift from that at low redshift.

For example, a host galaxy that has undergone many generations of star formation will have built up a higher density of the heavy elements (the astronomers call this "metallicity"), and one might imagine that this might lead to a supernova explosion of differing brightness. The key point, however, is that different galaxies have begun their life at different times in history, so at any given redshift there will be a wide distribution of galaxy ages, and hence metallicities. The demographics may shift as we go back in redshift, such that the peak of the host galaxy age distribution becomes a little bit younger, but there are still examples of both young and old host-galaxy environments even in the nearby supernova population.

We therefore can study supernova evolution effects simply by looking at nearby supernovae across a wide range of host galaxy ages. For the relatively small samples that have already been studied, the light curve width-luminosity relation appears to account for any evolutionary differences quite well, as we have seen. However, we would like to be able to examine hundreds of nearby supernovae to find even small departures from, and refinements of, this calibration relation.

(Excerpt end)

Observation:

References to metallicity reveal a search for causes of a disruption in the expected fusion progression.
Clearly the mechanism for a supernova is not certain, with two "accepted simplified theories."
Galaxies could have stars with a composition slightly different than in another galaxy, let alone within the same galaxy. The stellar composition is expected to affect the supernova.

After a review of the listed supernova IA detections, recognized as "the relatively small sample" and with the recognition of the importance of the stellar composition before the event, the supernova behavior is not as predictable as claimed.

The first book used several excerpts from the 1998 study to show the study's conclusions were invalid.

## 6.1 Crab Nebula

There are 2 images, first is in optical; second is in X-ray.

Crab Nebula (or Messier 1 or M1) is a well known supernova remnant.
The Chinese recorded this explosive event in the year 1054. One could expect all that remains after over 1000 years is only cold debris.
On the contrary, the debris "field" is quite active electrically. When looking at the visual image of M1, the first impression of what is claimed to be a dust cloud is:

This is definitely not from an explosion in space!

In space, an explosion puts debris in motion and the vacuum cannot slow it down.
Either the debris should be gone or very dispersed.
This image is just wrong for what should be there after 1000 years.
With a second impression, many filaments are clearly visible. Uncharged dust particles could clump by gravity.

Many filaments have formed so clearly this cloud is plasma. The inner filaments are brighter. This supposed dust cloud is electrically active.

The outer edges of the cloud are not well defined but the cloud has a defined extent where the cloud roughly ends.

This cloud has structure and is not debris from an explosion

The second image is from Chandra Observatory in X-ray. Everything in X-ray is synchrotron radiation.

X-rays are very high frequency electromagnetic radiation but they have a simple, well known mechanism.

Here is a short quote from the European Synchrotron Radiation Facility web site:

"The entire world of synchrotron science depends on one physical phenomenon: When a moving electron changes direction, it emits energy. When the electron is moving fast enough, the emitted energy is at X-ray wavelength."

In a synchrotron, a controlled magnetic field changes the current's direction resulting in electromagnetic radiation.

One very striking feature in the X-ray image of M1 is a bright ring. There is a sphere at its center.

This ring does not belong here after an explosion. The ring even has beads! This must be a luminous plasma filament where the plasma flow in the filament is confined by its magnetic field.

The object at the center of the ring is also interesting with an apparent jet of material that begins roughly perpendicular to the ring and continues below into a less organized cloud which is very bright in X-ray.

The jet along the object's axis suggests this central object is a plasmoid which has opposing jets and was imaged at the core of the galaxy M87.

There is a possible second filamentary ring outside the inner ring. The outer ring is more defined where below the core but less defined to the right. To the upper right is an active cloud where it could be the result of a jet in the opposite direction from the other jet.

The core of M1 is clearly a plasmoid.

One conjecture is the 1054 supernova was a collapse of the plasmoid and several rings were generated as a result.

Crab Nebula was in the recent news.

The story's headline in Science News on June 24, 2019 was:

The highest-energy photons ever seen hail from the Crab Nebula

The news story is sensational proposing shock waves as the cause but the bottom line is the photon was a very short wavelength of electromagnetic radiation in the extreme gamma ray range.

There is important background needed for this news story and this plasmoid.

Excerpt from Wikipedia:

The crab nebula is in the Perseus Arm of the Milky Way galaxy, at a distance of 6,500 light-years from Earth. It has a diameter of 11 light-years and is expanding at a rate of about 1,500 kilometres per second or 0.5% of the speed of light.

Observation:
This velocity is probably obtained from the spectrum of the plasma filaments. That measurement is not necessarily the velocity of this "dust cloud."

(Excerpt continued)

At the center of the nebula lies the Crab Pulsar, a neutron star with a spin rate of 30 times per second, which emits pulses of radiation from gamma rays to radio waves. At X-ray and gamma ray energies above 30 keV, the Crab Nebula is generally the brightest persistent source in the sky, with measured flux extending to above 10 TeV.
(Excerpt end)

There is a dramatic change from 10 to 100 trillion electron volts with a factor of 10 but the story omits that history, by just providing a huge number.

Every pulsar is known to be a plasmoid, in electric cosmology. Neutron stars are impossible and don't exist.

This particular plasmoid residing in the Perseus arm is able to draw on an energy source from its vicinity to increase its typical synchrotron radiation by a factor of 10. Even when at "rest" at 10 Tev this plasmoid is energetic.

Astronomers lack a long history of gamma ray observations so it is impossible to know the significance of the June energy burst. They apparently fail to recognize gamma rays are at the high frequency end of synchrotron radiation.

Astronomers are apparently not reacting to this activity in the crab nebula which should be impossible after a catastrophic explosion.

Apparently this supernova is not what it is claimed to be.

Hannes Alfven had observed extreme electrical events which suggested the same could happen as astrophysical events.

The Crab Nebula suggests that explanation as well.

Links in References, for both optical and X-ray

## 6.2 SMC Supernova E0102

Chandra had a story titled:

E0102-72.3: Astronomers Spot a Distant and Lonely Neutron Star

This supernova follows the Crab Nebula description and its plasmoid in the above section.

E0102, located in the SMC, is also claimed to be a supernova remnant. There are similarities in the two supernova observations.

Excerpt from Chandra:

Chandra observations of E0102 show that the supernova remnant is dominated by a large ring-shaped structure in X-rays, associated with the blast wave of the supernova. The new MUSE data revealed a smaller ring of gas that is expanding more slowly than the blast wave. At the center of this ring is a point-like source of X-rays. Together, the small ring and point source act like a celestial bull's eye.

The combined Chandra and MUSE data suggest that this source is an isolated neutron star, created in the supernova explosion about two millennia ago. The X-ray energy signature, or "spectrum," of this source is very similar to that of the neutron stars located at the center of two other famous oxygen-rich supernova remnants: Cassiopeia A (Cas A) and Puppis A. These two neutron stars also do not have companion stars.

(Excerpt end)

Observation:

The page (link in References) offers views in different wavelengths.

The image in X-ray is important. X-rays are the result of high energy synchrotron radiation so a very strong electrical current is required to achieve the energy in X-rays.

In an electric cosmology, every neutron star is a plasmoid. A neutron star is impossible, a hypothesis never tested. This was covered in the earlier books.

In E0102, the entire outer ring is active in X-ray. E0102 also has prominent filament spokes between the plasmoid at the center and the outer ring indicating an electrical connection.

Excerpt continued:

Oxygen-rich supernova remnants like E0102 are important for understanding how massive stars fuse lighter elements into heavier ones before they explode. Seen up to a few thousand years after the original explosion, oxygen-rich remnants contain the debris ejected from the dead star's interior. This debris (visible as a green filamentary structure in the combined image) is observed today hurtling through space after being expelled at millions of miles per hour.

Observation:

No spectra are provided here but a paper on the Crab Nebula stated the spectra from the filaments provided the velocity values for the M1 motions.

An electrical current within a filament is not like molecules moving in a gas cloud. This extreme velocity is probably not justified.

Excerpt continued:

But how did this neutron star end up in its current position, seemingly offset from the center of the circular shell of X-ray emission produced by the blast wave of the supernova? One possibility is that the supernova explosion did occur near the middle of the remnant, but the neutron star was kicked away from the site in an asymmetric explosion, at a high speed of about two million miles per hour. However, in this scenario, it is difficult to explain why the neutron star is, today, so neatly encircled by the recently discovered ring of gas seen at optical wavelengths.
(Excerpt end)

Observation:

The assumption is only an explosion caused what is observed.
The lack of symmetry is awkward to explain because the star which exploded with its "blast wave" should have been a symmetrical sphere. The article describes possible scenarios for the observation but without a conclusion.

An electrical collapse is not required to maintain symmetry with the filamentary structures which emerged from the event.

The oxygen content in the debris is assumed to be a clue about the star's core at the time of the supernova.

The SAFIRE project has confirmed a number of elements are created on the anode's surface. Our Sun's surface has most of the periodic table, observed by their absorption lines, suggesting an extreme age.

Any conclusions drawn from the metallicity in the debris about the original star are invalid.

Conclusion:

Cosmologists have more to learn about a supernova. They are rare. Unfortunately it is impossible to know what was present before this E0102 supernova event which resulted in the observed debris.

With the solar model based on liquid metallic hydrogen, a supernova is the ejection of the star's outer layer, the photosphere, which is composed of liquid metallic hydrogen, which is a lattice of protons maintained by electrons. Ejecting the photosphere is the ejection of a sphere of plasma which at the time was conducting an electric current.

After ejection, the plasma would form filaments simply because charged particles in motion form filaments where the magnetic field generated by charges in motion constricts the flow; that is why it is called plasma.

## 6.2 Cassiopeia A is a Supernova

Chandra had a story titled:

Chandra Reveals the Elementary Nature of Cassiopeia A

This section follows other supernovae including the Crab Nebula and E0102, both being a supernova and leaving a neutron star, or in an electric cosmology, a plasmoid.

Cassiopeia A was a bright star at this location recorded in 1630 and in 1680. With that double brightening perhaps this is not a typical supernova.

From Wikipedia: "At any rate, no supernova occurring within the Milky Way has been visible to the naked eye from Earth since [1680]."

In 1964 it was not observed by a rocket checking for X-ray sources.
In 1965 another rocket detected it in X-ray; this source is called Cas X-1 or Cas A.

Excerpt from Wikipedia:

[As a radio source] Cas A had a flux density of 2720 ± 50 Jy at 1 GHz in 1980 Because the supernova remnant is cooling, its flux density is decreasing. At 1 GHz, its flux density is decreasing at a rate of 0.97 ± 0.04 percent per year. This decrease means that, at frequencies below 1 GHz, Cas A is now less intense than Cygnus A. Cas A is still the brightest extrasolar radio source in the sky at frequencies above 1 GHz.

In 1999, the Chandra X-Ray Observatory found a "hot point-like source" close to the center of the nebula that is the neutron star remnant left by the explosion.

In 2013, astronomers detected phosphorus in Cassiopeia A, which confirmed that this element is produced in supernovae through supernova nucleosynthesis. The phosphorus-to-iron ratio in material from the supernova remnant could be up to 100 times higher than in the Milky Way in general.

(Excerpt end)

Observation:

With the E0102 supernova, the presence of oxygen is assumed to be from only the star's interior and not a product of the supernova. Here, phosphorus was a product of the event.

Excerpt from Chandra:

The exact means by which a massive explosion is produced after the implosion is complicated, and a subject of intense study, but eventually the infalling material outside the neutron star was transformed by further nuclear reactions as it was expelled outward by the supernova explosion.

Astronomers have found other elements in Cas A in addition to the ones shown in this new Chandra image. Carbon, nitrogen, phosphorus and hydrogen have also been detected using various telescopes that observe different parts of the electromagnetic spectrum.

Combined with the detection of oxygen, this means all of the elements needed to make DNA, the molecule that carries genetic information, are found in Cas A.

(Excerpt end)

Observation:
The article has much detail about the assumptions of a star's fusion cycle and specific elements.

There are also interesting assumptions about stellar changes before the main event, with the star beginning at a mass 16 times the Sun but somehow "[it] lost roughly two-thirds of this mass in a vigorous wind blowing off the star several hundred thousand years before the explosion" "[so] the doomed star was about five times the mass of the Sun just before it exploded."

These proposed changes in mass have no observed evidence to justify them. The source of the force needed for the "vigorous wind" which removes many solar masses is not identified.

The actual date of the supernova is not clear. Its first bright phase might be in 1680. It was first observed in X-ray in 1965, but in the year before, its location was absent in X-ray.
The date when this debris field was generated is ambiguous.
The 3-D explosion simulation has a 330-year span for a start around 1680 implying both 1964 and 1965 were long after the explosion which is awkward timing for the start of X-ray emissions.

Near the end of the article is this:

"The different datasets have revealed new information about the neutron star in Cas A, the details of the explosion, and specifics of how the debris is ejected into space."

Somehow, they have determined the "explosion" sequence which required a "detonation" mechanism which was never observed, just from a few changes in brightness and the observed nebula.

In a new cosmology, a neutron star is a plasmoid.

There are separate links in that web page statement to the details about the neutron star and details about the debris. Each link will be mentioned below.
Partial excerpt from the neutron star addition:

This new research has allowed the teams to place the first observational constraints on a range of properties of superfluid material in neutron stars. The critical temperature was constrained to between one half a billion to just under a billion degrees Celsius. A wide region of the neutron star is expected to be forming a neutron superfluid as observed now, and to fully explain the rapid cooling, the protons in the neutron star must have formed a superfluid even earlier after the explosion. Because they are charged particles, the protons also form a superconductor.

Using a model that has been constrained by the Chandra observations, the future behavior of the neutron star has been predicted. The rapid cooling is expected to continue for a few decades and then it should slow down.
(Excerpt end)

Observation:
These false extreme temperatures come from the plasmoid's synchrotron radiation in X-ray.

X-ray is never in thermal radiation and can never represent a temperature.

These physicists are busy developing a model for a non-existent neutron star while observing a plasmoid and measuring its plasma behaviors, which are assumed to be like a "very hot gas."

Another excerpt:
"These neutrinos escape from the star, taking energy with them and causing the star to cool much more rapidly."

Observation:
Proposing a transfer of thermal energy into kinetic energy carried by neutrinos could be a new theory with no observed evidence. However the stability of a huge entity consisting of only neutrons with no protons and having a strong magnetic field s also unverified, or impossible.

This story included another new theory: "Evidence for a bizarre state of matter - known as a superfluid - has been found in Cassiopeia A."

If it is "bizarre" then the conclusion probably has no precedence in valid physics. This mention of neutrinos for cooling is also bizarre.
A plasmoid is electrically driven so a reduction in its electric current results in a reduction in its radiation. There is nothing bizarre here.

Excerpt from the debris addition:
Scientists have combined data from Chandra, NASA's Spitzer Space Telescope, and ground-based facilities to construct a unique 3D model of the 300-year old remains of a stellar explosion that blew a massive star apart, sending the stellar debris rushing into space at millions of miles per hour.

(Excerpt end)
Observation:

This "debris" page is worthwhile for the X-ray image of the 'debris field' which is roughly a sphere.

The entire field is dominated by filaments, even the entire circumference.

From the image, one cannot tell if this is only a hollow shell or whether there are any plasma filaments inside, as seen to the ring around E0102. Unlike E0102, the debris field for Cas A looks symmetrical.

The "debris field" for each supernova is dominated by plasma filaments.
This supernova is an active electrical entity, not just remnants of a catastrophic explosion.

The superfluid reference suggests liquid metallic hydrogen. Which is the basis of is the solar model proposed by Crothers and Robitaille in the Electric Universe cosmology.

Perhaps there is another interpretation of Cassiopeia A:

a)

The remnant at the core is the original star, composed of condensed matter. This star as condensed matter is dense but certainly not just neutrons. Metallic hydrogen is a lattice of protons whose structure is maintained by electrons.

What cosmologists call a neutron star is a condensed matter star, I made a mistake to follow the distracting claim there is a neutron star here. The original star was and is a conventional star, never a neutron star.

After the ejection of the photosphere, the star's core and convective zone switch to an electric discharge mode which is dominated by synchrotron radiation, over the thermal radiation from the stellar remnant.

b)
There is a huge plasma cloud near this central star. The cloud is a plasma entity with many filaments maintaining its integrity.

Any plasma filaments between remnant and cloud are the maintained electrical connections between them.

## 6.3 Supernova verify

A supernova is considered a candidate for a standard candle, which is an accurate mechanism for calculating a distance by its light dimmed in a predictable manner by the distance to its source.

After some research, there is no basis for this expectation of a supernova.
Claims of an observed supernova must be verified to establish their validity. If it is just a variable star, then it must be treated as one of that type. There are many of these with different star types and different luminosity curves.

This section is long because it has much data to justify the conclusion.

Astronomers have tried to explain this mechanism in a star which results in the observed brightening assumed to be an explosion.
NASA has an Objects of Interest page titled: Supernovae
Excerpt:

Supernovae are divided into two basic physical types:
Type Ia. These result from some binary star systems in which a carbon-oxygen white dwarf is accreting matter from a companion. (What kind of companion star is best suited to produce Type Ia supernovae is hotly debated.) In a popular scenario, so much mass piles up on the white dwarf that its core reaches a critical density of $2 \times 10^9$ g/cm$^3$. This is enough to result in an uncontrolled fusion of carbon and oxygen, thus detonating the star.

Type II. These supernovae occur at the end of a massive star's lifetime, when its nuclear fuel is exhausted and it is no longer supported by the release of nuclear energy. If the star's iron core is massive enough, it will collapse and become a supernova.
However, these types of supernovae were originally classified based on the existence of hydrogen spectral lines: Type Ia spectra do not show hydrogen lines, while Type II spectra do.

Gravity gives the supernova its energy. For Type II supernovae, mass flows into the core by the continued formation of iron from nuclear fusion. Once the core has gained so much mass that it cannot withstand its own weight, the core implodes. This implosion can usually be brought to a halt by neutrons, the only things in nature that can stop such a gravitational collapse. Even neutrons sometimes fail depending on the mass of the star's core. When the collapse is abruptly stopped by the neutrons, matter bounces off the hard iron core, thus turning the implosion into an explosion.
For a Type Ia supernova, the energy comes from the runaway fusion of carbon and oxygen in the core of a white dwarf.
(Excerpt end)

Definitions from Wikipedia topic for Supernova:

Type Ia has singly ionized Silicon (S II) absorption line at peak light. Cause is "thermal runaway."

Type Ib shows non-ionized helium (He I) absorption line.
Type Ic has weak or no helium absorption line.
Type II-P reaches "plateau" in its light curve.
Type II-L has "linear" decrease in light curve (linear in magnitude versus time)

Cause of all types except type Ia is a "core collapse."

Observation:
Type IA definition has settled, after being "hotly debated" for a binary pair combination, on only a "popular scenario" with 'uncontrolled fusion" for an explanation.

This ill-defined scenario cannot be a good candidate. If uncontrolled it is not predictable.

Type II has gravitational collapse causing an explosion as the explanation.
This ill-defined scenario cannot be a good candidate.

Observation:
Astronomers lack an understanding of a supernova. A sequence of brightening cannot be explained to provide a prediction.

Without a valid explanation of the process, the definition of a supernova is not worthy of "super" but rather is just a nova, a star whose brightness changes due to a natural process, rather than an explosion.

The nova is also unexplained but this section is about a supernova.

These supernova scenarios are not predictable for use as a standard candle.


Despite lacking a basis, astronomers have observed many stars which brightened, called many a supernova and tried to identify a type for that explosive brightening.

Supernova survey systems have been developed to catch a number of "new" bright stars during the continuous sweep of the sky by automated digital comparisons of images.

The following is most supernovae since roughly 1800. The list ends here when the events become too dim (to reduce the list).

From Wikipedia:
"no supernova occurring within the Milky Way has been visible to the naked eye from Earth since [1680]."

Observation:

The problem with each remote supernova is the host galaxy distance will be wrong (by its redshift miscalculation) resulting in a wrong brightness expectation to start with, when the baseline is a star in the Milky Way.

Data from Wikipedia list of supernova:

MM or maximum magnitude is from the Open Supernova Catalog when available. The star Vega is 0; 6 is visible to the naked eye; the increasing value is dimmer.

SN1885A was in M31, at 2.4Mly, but is type I peculiar. MM=9.0

SN1937C was in IC 4182 at 13Mly, is type Ia. MM=13.96
SN1940B was in NGC 4725 at 36Mly, is type ii-P. MM=12.8
SN1940A was in NGC 5907 at 50Mly, is type ii-P. MM=14.33
SN1940C was in IC 1099 (no d), is type ii-P. MM=16.3

SN1972E was in IC 3253 at 11Mly, is type Ia. This is considered the protype Ia. MM=7.77
SN1983N was in M83 at 15Mly, is type Ib, the first Ib.
SN1987A was in LMC at 160kly, is type II-peculiar. MM=1.9
SN1993J was in M81 at 11Mly, is type IIb. MM=9.91
SN1994D was in NGC 4526 at 50Mly, is type Ia.
SN1998bw was in ESO 184-G82 at 140Mly, is type Ic.
SN2002bj was in NGC 1821 at 160Mly, is type Ia.
SN2003fg was in unnamed at 4Mly, is type Ia. (no basis for distance)
SN2004dj was in NGC 2403 at 8Mly, is type Ia(SAB).
SN2005ap was in unknown at 4.7By, is type II. MM=12.04 (no basis for distance)

SN2005gj was in NGC 266 at 200Mly, is type II-n for a new type.

240SN2006gy was in NGC 160 at 240Mly, is type IIn.
SN2007bi was in unnamed at 160Mly, is type Ia.
SN2007uy was in NGC 2770 at 84Mly, is type Ibc.
SN2008D was in NGC 2770 at 88Mly, is type Ibc; claimed to be observed while exploding just by its changes in brightness.
Wikipedia has NGC 1770 at different distances.

in 2009: MENeaC was in globular cluster of unnamed galaxy in a glaxy cluster at 1,000Mly, is rated type Ia.(no basis for a distance; it is considered "associated" with a globular cluster.)

SN2010lt was in NGC 3378 at 240Mly, is type Ia(sub-luminous). (it reached only magnitude +17 so it was very faint; but was type Ia??)

SN2011fe was in M101 at 21Mly, is type Ia. MM=9.48
SN2014j was in M82 at 11.5Mly, is type Ia. MM=8.89

From the Open Supernova Catalog in addition to those above; OSC does not list a distance:

Sorted by date:
SN1895B is type Ia. MM=7.07
SN1954A is type Ia. MM=9.1

SN1986G is type Ia. MM=10.61
SN1994D is type Ia. MM=10.96
SN1999da is type Ia. MM=0.13
SN1999dk is type Ia. MM=0.086
SN1999gp is rated IIb. MM=0.006
SN2000ce is rated Ia. MM=0.272

The last few are the brightest in the list.

More sorted by date, but dimmer than above:
SN1989B is type Ia. MM=11.2
SN1962M is type II. MM=11.4
SN1998bu is type Ia. MM=11.44
SN1999ee is type Ia. MM=12.8
SN1999em is type II-P. MM=12.8
SN2000cx is type Ia-Pec. MM=12.9
SN2002ap is type Ic. MM=12.04
SN2003dh is type Ia. MM=12.62
SN2004ef is type Ia. MM=10.32
SN2005cf is type Ia. MM=13.32
SN2006X is type Ia. MM=12.53
SN2007af is type Ia. MM=13.05
SN2008ax is type IIb. MM=12.07
SN2009dc is type Ia-Pec. MM=14.53
SN2009ip is type IIn. MM=12.01
SN2010jl is type IIn. MM=10.62
SN2011dh is type Ic. MM=12.62
SN2012dn is type Ia. MM=13.67
SN2012ap is rated type Ic. MM=12.04
SN2012aw is type II-P. MM=11.96
SN2012ht is type Ia. MM=12.91
SN2012r is type Ia. MM=11.74
SN2013dy is type Ia. MM=12.8
SN2013aa is type Ia. MM=11.28
SN2013ej type II-P/L. MM=11.52
in 2013: IPTF13bvn is type Ib. MM=14.75
SN2014J is type II-P/L. MM=8.98
in 2014:ASASSN-14ha is type II. MM=12.23

there are more with MM > 11.4

For comparison:
Object and magnitude
SN1006 -0.8 - recorded around the world
SN1054 -4.0 - recorded in China, Japan, Arabia; is now the Crab Nebula
SN1572 -4.0 - Tycho's Nova
SN1604 +2.95 - Kepler's Supernova

Ceres +6.64
Iapetus +10.20
Phobos +11.30
Diemos +12.40
Pluto +13.65
Eris +18.7

SN1972E +7.77 is is considered the supernova prototype for type Ia

SN2010lt +17.0 is type Ia supernova though dim

Observation:

1) Cosmologists do not have a complete explanation of a supernova event.

Without that, each brightening could be anything else.
2) They rely on only several absorption lines, helium and silicon. Both are in the Sun though Silicon is only 0.002% but present.

3) These supernova types with inadequate detail cannot provide an accurate benchmark brightness to serve as a standard candle.

4) Most supernova events since 1680 are dim.

These are claimed to be catastrophic events but none are easily visible.

This suggests a real supernova explosion is as rare as a millennium.

5) Many appear to be just a nova, or perhaps a variable star with a long period, not a supernova which one could expect should be much brighter than either a nova or a periodic variable star.

This spectrum analysis of absorption lines seems inadequate to distinguish between types for the supernova explosion.
Observed evidence of an explosion is required to verify the selection of an exceptional claim of the unusual supernova.

6) Most are dim so they must be distant regardless of the distance assumed for the host galaxy. The accuracy for comparing the benchmark to a dim object might have a significant margin of error.

The critical criteria for a standard candle are accuracy and consistency.

Observation:

Only a very small number of actual supernova events occur when a star becomes unusually bright. These include the 4 bright events listed above. All 4 have something visible as a remnant to confirm the brightening was unusual.
The evidence must be provided for each claim. These 4 confirm no others.

Every one of these supernovae having an assigned type must be imaged in X-ray to observe evidence for an explosion.

SN2008D is observed in X-ray to fluctuate. That observation looks like a nova with no debris.

Until that verification is done:

Each event is just a nova or just an intermittent change in brightness, like SN2008D.

The obvious candidate needing verification is SN2010lt; this object was declared a type Ia supernova despite it being only slightly brighter than Eris, a distant TNO!

We are given the unjustified impression astronomers are detecting supernovae.

Unless evidence is provided, many claimed supernovae are probably just nova events. Whatever light curve is observed is not from an explosion.

Until the observed variation in a nova has been suitably explained to be predictable, a variable star like a nova cannot be used as a standard candle.

Just because we can see the Crab Nebula does not mean astronomers know what a supernova actually is.

## 6.4 Nova or Supernova

This section is a follow up to the section above concerning a supernova without verification is probably just a variable star.

However there are actually several variable star events which might be called a supernova given the weak definitions provided for these events.

There are several types of a nova event.

Excerpt from Wikipedia:

A nova is a transient astronomical event that causes the sudden appearance of a bright, apparently "new" star, that slowly fades over several weeks or many months. Causes of the dramatic appearance of a nova vary, depending on the circumstances of the two progenitor stars. All observed novae involve a white dwarf in a close binary system. The main sub-classes of novae are classical novae, recurrent novae (RNe), and dwarf novae. They are all considered to be cataclysmic variable stars.

Classical nova eruptions are the most common type of nova. They are likely created in a close binary star system consisting of a white dwarf and either a main sequence, subgiant, or red giant star. When the orbital period falls in the range of several days to one day, the white dwarf is close enough to its companion star to start drawing accreted matter onto the surface of the white dwarf, which creates a dense but shallow atmosphere. This atmosphere, mostly consisting of hydrogen, is thermally heated by the hot white dwarf and eventually reaches a critical temperature causing ignition of rapid runaway fusion.

The sudden increase in energy expels the atmosphere into interstellar space creating the envelope seen as visible light during the nova event and previously mistaken as a "new" star. A few novae produce short-lived nova remnants, lasting for perhaps several centuries. Recurrent nova processes are the same as the classical nova, except that the fusion ignition may be repetitive because the companion star can again feed the dense atmosphere of the white dwarf.

Novae are classified according to the light curve development speed, so in

NA: fast novae, with a rapid brightness increase, followed by a brightness decline of 3 magnitudes — to about 1/16 brightness — within 100 days.
NB: slow novae, with magnitudes of 3, decline in 150 days or more.
NC: very slow novae, also known as symbiotic novae, staying at maximum light for a decade or more and then fading very slowly.
NR/RN: recurrent novae, novae with two or more eruptions separated by 10–80 years have been observed.

Some novae leave behind visible nebulosity, material expelled in the nova explosion or in multiple explosions.

(Excerpt end)

Observation:

The supernova type Ia had been "hotly debated" whether it was a white dwarf event but "thermal runaway" was the consensus. The "classical nova" with its description of 'critical temperature causing ignition of" rapid runaway fusion" is rather similar to a supernova type Ia.

The peak magnitude is never defined for the novae. The light curves use a range indicating these events are not consistent.

Not only do astronomers fail at explaining a supernova, they fail at explaining the nova as well when using a similar ill-defined mechanism

If anything is in a "runaway" mode then it must be in violation of thermodynamics and its conservation of energy. A system cannot create more heat when the internal system cannot (like from an exothermic chemical reaction)

This lack of distinction for a nova gets worse when looking at some variable star types.

From the list of variable star types:

"Dwarf novae are stars involving a white dwarf in which matter transfer between the component gives rise to regular outbursts."

This 'variable star' type of a "dwarf nova" is much the same as others described above.

I suspect the fusion model being wrong results in many wrong explanations for any observed change in a star's brightness.

Many of these descriptions typically falter at either explosion or runaway.

One could hope a behavior like variability having an observed range can be explained without these extreme effects beyond a definable process conforming to physics.

The white dwarf is required for many explanations so this specific star type is critical in these particular behaviors in cosmology.

Perhaps only the variable star type of an eclipsing binary is not affected by the model used for the star.

Neither the nova nor supernova has a clear explanation for the observed variability.

## 6.5 SN1979C Supernova to X-ray Source

M100 Galaxy had a recent supernova SN 1979C. That now faint object is a strong X-ray source, so of course it is claimed to be a black hole.

Chandra had a story titled:

SN 1979C: NASA's Chandra Finds Youngest Nearby Black Hole

From Wikipedia on SN 1979C:

On November 15, 2010 [31 years later] NASA announced that evidence of a black hole had been detected as a remnant of the supernova explosion.

The researchers observed a steady source of X-rays and determined that it was likely that this was material being fed into the object either from the supernova or a binary companion. However, an alternative explanation would be that the X-ray emissions could be from the pulsar wind nebula from a rapidly spinning pulsar, similar to the one in the center of the Crab Nebula. These two ideas account for several types of known X-ray sources. In the case of black holes the material that falls into the black hole emits the X-rays and not the black hole itself. Gas is heated by the fall into the strong gravitational field.
SN 1979C has also been studied in the radio frequency spectrum. A light curve study was performed between 1985 and 1990 using the Very Large Array radio telescope in New Mexico.

(Excerpt end)

Observation:

One would expect the supernova would scatter its debris far from the super explosion.

Somehow:

1) The remnant of a star, apparently having so much remaining mass, just gave up being a star and collapsed into a black hole.

2) While collapsing into this impossible "black" hole in a non-existent remote moving observer's space-time,
3) It also gathered up that far-flung, departing, debris field,

4) Compressed that debris into an accretion disk,
5) Heating by only friction, to an impossible temperature,

6) Becoming bright in X-ray,
7) And in radio which is never in thermal radiation (radio is only in synchrotron radiation),

8) Within only 31 years.

That part of this story is just silly nonsense. Black holes don't exist.

The X-ray point source accompanied by must be a plasmoid with its broad frequency range of synchrotron radiation covering from X-ray to radio.

What is also interesting in this story is the set of Chandra images in different wavelengths of M100 and its periphery.

Excerpt from Wikipedia:

Messier 100 (also known as NGC 4321) is a grand design intermediate spiral galaxy. Messier 100 is considered a starburst galaxy with the strongest star formation activity concentrated in its center, within a ring - actually two tightly wound spiral arms attached to a small nuclear bar with a radius of 1 kilo-parsec – where star formation has been taking place since at least 500 million years ago in separate bursts.

Excerpt end)

Observation:

M100 is classified as type SAB(s)bc, not explicitly a Seyfert, though it is called a starburst galaxy.

From Wikipedia:
Seyfert galaxies are one of the two largest groups of active galaxies, along with quasars. They have quasar-like nuclei (very luminous, distant and bright sources of electromagnetic radiation) with very high surface brightnesses whose spectra reveal strong, high-ionisation emission lines, but unlike quasars, their host galaxies are clearly detectable.

(Excerpt end)

Observation:

Seyfert galaxies were associated with quasars by Halton Arp's many observations noted in his book Seeing Red. If the M100 spectrum were provided maybe we could help decide on the classification as Seyfert or not...

References links provides the page with the Chandra story and its images.

Clicking on the X-ray tab shows that wavelength image.

Moving the cursor over this image pops up the pointer to SN 1979C at the lower left.

About a dozen X-ray sources can be counted, scattered about this "starburst" galaxy.

M100 is literally bursting with plasmoids!

Clicking on Infrared or Optical reveals the plasmoids dim away at longer wavelengths, but not short wave lengths like X-ray. Quasars are typically dimmed when shrouded in clouds of metallic ions so this observation suggests perhaps some of the point sources are actually quasars. No spectra are provided to confirm this conclusion.

M100 follows the pattern of the second book, Cosmology Transition, where an active spiral galaxy (including M51 and M82) is surrounded by scattered X-ray point sources, which are plasmoids not black holes.

Observation:

One can only wonder what caused SN 1979c to brighten then dim in optical but remain bright in X-ray. Quasars are shrouded in clouds of metallic atoms so are often dimmed in optical but not in X-ray. If the cloud of ions dispersed, the optical would brighten with no change in X-ray.

In a laboratory, plasmoids do not persist a long time. On the cosmological scale, they persist much longer.

A quasar spectrum would certainly be interesting for any of the dozen, including the SN.

The Wikipedia topic does not mention whether anyone found an image with a star before this supernova.

Scenarios:
a) the plasmoid arose at that spot, or
b) the plasmoid moved to there.

If (a) how?

if (b) how fast and why did it stop?

The same 2 questions apply to the other scattered X-ray sources below the galaxy. several sources are in the spiral arms and there is an opposing pair at 4 and 10 o'clock to the core, so this dozen has no clear pattern.

Having no spectra, one can only wonder, when lacking spectra, whether M100 actually ejected a quasar...

# 7 Plasmoid

Excerpt from Wikipedia:

A plasmoid is a coherent structure of plasma and magnetic fields. Plasmoids have been proposed to explain natural phenomena such as ball lightning, magnetic bubbles in the magnetosphere, and objects in cometary tails, in the solar wind, in the solar atmosphere, and in the heliospheric current sheet. Plasmoids produced in the laboratory include field-reversed configurations, spheromaks, and in dense plasma focuses.
The word plasmoid was coined in 1956 by Winston H. Bostick to mean a "plasma-magnetic entity":
The plasma is emitted not as an amorphous blob, but in the form of a torus. We shall take the liberty of calling this toroidal structure a plasmoid, a word which means plasma-magnetic entity. The word plasmoid will be employed as a generic term for all plasma-magnetic entities.

(Excerpt end)

Thunderbolts Project has a YouTube video clearly explaining a plasmoid, titled:

## 7.1 Black Hole or Plasmoid?

A black hole does not exist but one is proposed for the M87 galaxy core.

This video explains a plasmoid and the object observed in M87, in April 2020. Its title:

Thornhill: Black Hole or Plasmoid? | Space News

## 7.2 NGC 4194: A Black Hole in Medusa's Hair

That is the title of Chandra's story.

The page offers selecting separate images for different X-ray wave lengths.

Excerpt:

This composite image of the Medusa galaxy (also known as NGC 4194) shows X-ray data from NASA's Chandra X-ray Observatory in blue and optical light from the Hubble Space Telescope in orange. Located above the center of the galaxy and seen in the optical data, the "hair" of the Medusa -- made of snakes in the Greek myth -- is a tidal tail formed by a collision between galaxies. The bright X-ray source found towards the left side of Medusa's hair is a black hole (rollover the image to view).
Most bright X-ray sources in galaxies are binaries containing either stellar mass black holes or neutron stars that remain after the supernova explosion of a massive star. Because these compact objects can generate X-rays for much longer periods of time than the lifetime of their massive progenitor stars, X-ray binaries may be used as "fossils" to study the star formation history of their host galaxies. In this Medusa image, the X-ray binaries are seen as the bright blue point-like objects.
A recent study of the Medusa galaxy and nine other galaxies measured the correlation between the formation of stars and the production of X- ray binaries. A key feature was to study this correlation for the Medusa galaxy and NGC 7541, two galaxies with particularly high star formation rates. It was found that both the number of bright X-ray sources and their average brightness were related to the rate at which stars formed.

This work may be useful for attempts to use X-ray brightness to measure the rate of star formation in galaxies at very large distances.

(Excerpt end)

Observation:

This story is about ejection of plasmoids. The image reveals X-ray activity, from energetic plasma. The "cloud" of X-ray emissions around the galaxy core is energetic plasma.

When high velocity charged particles, like electrons and protons, change their path by a magnetic field, synchrotron radiation results. The peak frequency is driven by the velocity of the particles. Their velocity here is high enough for X-ray energy. The "cloud" implies a diffused flow like many particles in spiral paths, which is a diffused source.

## 7.3 M82: Chandra Images Torrent of Star Formation

That is the title of a Chandra story.

There are two separate news stories about M82 using Chandra images but the two stories are remarkable in comparison.

The first Chandra story (2011) has a wide angle image around M82.

Excerpt:

This new deep Chandra image reveals hundreds of point-like X-ray sources, some of which likely contain black holes.

Supernova explosions have produced bubbles of hot gas that extend millions of light years away from the plane of the galaxy.

(Excerpt end)

Observation:

I cannot recall any supernovae creating bubbles of hot gas that extend beyond their galaxy. Atoms in a gas can do only emission lines which I expect can never reach the X-ray frequency energy. However "hot gas" as plasma when the flow changes direction will emit synchrotron radiation. With a high velocity plasma flow, X-rays can be the peak frequency.

Cosmologists propose those distant sources as likely black holes but cosmologists have no alternate suggestion.

This image shows M82 galaxy literally in the middle of many scattered X-ray sources. There is no explanation offered for this observation. Instead, a "brush with M81 set off this torrent of star formation" in M82, which is a starburst galaxy. That "brush" refers to only stars forming in M82 but not those X-ray point sources out in the wide space around M82.

Here is a simple summary of these X-ray sources in the image:

There are several near the core, and 3 in a line which in optical are in the galactic disk, but the rest of the hundreds are outside the halo and disk.

Most of those well outside the core have the same intensity as those near the core implying these sources are the same type of object.

Noticing 3 of them are in the disk could suggest they are just a few unusual stars intense in X-ray. That coincidence fails with similar sources found both above and below the disk and the rest are in an almost random distribution in space.

The NASA story (2010) titled "Starburst Galaxy M82" emphasizes its Chandra X-ray image (with the story) which shows two possible black holes at the core but says nothing about the many sources surrounding the galaxy.

An excerpt from its caption:

The pullout is a Chandra image that shows the central region of the galaxy and contains two bright X-ray sources, identified in a labeled version.

(Excerpt end)

Observation:

Those 2 hide the many which are of special interest. This pullout is a zoom into the core but this pullout hides many of those sources surrounding M82. This placement could be either accidental or intentional. With the explicit "special interest" this placement seems intentional.

I did not select this NASA story and its image for the beginning because the core is not of special interest. The text offers much detail about the analysis of the black hole pair, not to be discussed here. Their "interest" is in multiple X-ray point sources around the core.

The reader of the NASA page gets the impression there are fewer here at M82 than a reader of the Chandra page.

Excerpt from Wikipedia:

In the core of M82, the active starburst region spans a diameter of 500 pc. Four high surface brightness regions or clumps (designated A, C, D, and E) are detectable in this region at visible wavelengths. These clumps correspond to known sources at X-ray, infrared, and radio frequencies.

The Chandra X-ray Observatory detected fluctuating X-ray emissions from a location approximately 600 light-years away from the center of M82. Astronomers have postulated that this fluctuating emission comes from the first known intermediate-mass black hole, of roughly 200 to 5000 solar masses. M82, like most galaxies, hosts a supermassive black hole at its center with a mass of approximately $3 \times 10^7$ solar masses as measured from stellar dynamics.

(Excerpt end)

A summary of the pullout image:

The core consists of one very bright source (X1), 3 sources (X3) in a pyramid to its left and another bright source (X2) at a distance in the 4 o'clock direction, having another dimmer source nearly adjacent to it.

An intermediate source (X4) is below the pyramid in the active cloud.

An intermediate source (X5) is below the core in the active cloud.

There are about 11 other slightly dimmer sources scattered about the 4 in their close bunch (called X1+X3 here).

This summary might have some similarities but I could find no image identifying the A, C, D, E clumps mentioned. The set of [ A=X1 + X3, C=X4, D=X5, E = intense cloud left of X3 ] is one possibility. This galaxy has distorted arms so perhaps some of that distortion is in the line of sight with the core. The core explanation is not the main topic of this description.

The pulsar is probably X2 but the image has no scale to check 600ly.

From the messier-objects-com site: "The pulsar in M82 was given the designation M82 X-2. Its luminosity is 100 times greater than its mass should be able to produce, in theory. The pulsar is classified as an ultraluminous X-ray source (ULX)."

This description will ignore the M82 pulsar. Thornhill has a video about neutron stars and pulsars, mentioned in the Neutron stars section.

One possible question for the M82 core: Could it resemble that of a spiral galaxy?

Excerpt from Wikipedia:

M82 was previously believed to be an irregular galaxy. In 2005, however, two symmetric spiral arms were discovered in near-infrared (NIR) images of M82.

(Excerpt end)

Observation:

If the galaxy is rotating then it should have a magnetic field around the electric current through its axis, like in the Milky Way.

If there is a Z pinch with a Birkelund filament pair in M82, there is no evidence in the images. However there is not a good X-ray image of the Milky Way core. There are images of its pairs of "bubbles" or "chimneys" but the Z pinch with its pair of electric currents is not an actual object.

Images of the Milky Way core reveal many stars in chaotic motion among clouds of gas and dust. The distribution in the M82 core looks chaotic.

The M82 core shows many lobes not just two.

Back to the first image with its own special interest ...

The big mystery is what are these many X-ray point sources? They look like stars but are far brighter than the normal stars in M82 and its normal stars not bright in X-ray.

The other known X-ray point sources (not a fictional black hole) are either

1) Synchrotron radiation from electrons changing direction in a magnetic field, but this combination is unlikely at a great distance from a galaxy, or

2) a plasmoid.

A plasmoid is the better selection because one was observed in M87 while stars which are very strong X-ray sources while at a remote distance from a galaxy are not observed with this unique exception of hundreds somewhat near M82 and no other galaxy.


The obvious mystery here is:

Why do they have the observed random distribution?

Quasars are often in pairs. That consistent pattern is not here.

Their formation out in intergalactic space is unlikely, so they must have moved there from wherever they formed. An undefined "brush by" event having no details is not even an inadequate proposal. It is just a worthless conjecture.

An ejection requires an undefined mechanism but modern cosmologist have readily proposed the random interactions among a small number of bodies held loosely by gravity can result in an ejection just by the small system's chaos.

I am not suggesting that here. I am only pointing out ejecting a body out of a system is a well known scenario for many contexts including our solar system when checking the limits for its long term stability after a perturbation.

Some candidates for ejections from the core, using the color image in the story, are possibly found:

at 12 o'clock is a separated pair, 1 solitary bright source above the cloud while at 6 o'clock is a bright source in the red cloud.

at 1 o'clock there are 2 in line with the first at the edge of the red cloud.

at 2 o'clock there are 3 with a pair at the edge and the third distant.

at 4 o'clock there are 3 in line, in the disk, with the last as the brightest.

at 5 o'clock there are 3 in line, with the first in the cloud.

at 6 o'clock there are 2 in line with the core, with the closer part of a pair.

at 10 o'clock there are 3 in line with the core, with the first as the brightest and the middle as red.

These 18 could be just coincidences among what seems a random pattern.

The other coincidence is all of the most distant sources have another source between them and the core.

I find the coincidences, contrived or not, worth noting. Randomness is not a good start.

Astronomers are obsessed with black holes in general but are unconcerned about many spread across a wide area surrounding M82.

The observation of hundreds of sources around M82 is significant and needs a viable explanation.

The big question, after discovering these images of M82 is: how common are strong X-ray point sources in the space far between galaxies??

I suspect the number is extremely low.

M82 is sometimes considered relatively close to M81 but their distance is very roughly 300 kly

M82 and its collection present a fascinating mystery.

Before a conclusion, another observation is necessary.

The Wikipedia topic for M82 provides a high resolution image.

One can observe there is material from the core in an 11 o'clock direction. There is also material in the opposite direction from the core. For both, the material has structure and is not an amorphous cloud. The red is probably hydrogen gas clouds found through atoms with their specific emission line while the brown is probably dust clouds. The image suggests these closer gas and dust clouds rotate with the stellar disk behind them.

M82 is a very interesting galaxy (an irregular type). Its periphery is as well.

# 7.4 Toothbush Cluster

The story title:

Featured Image: New Detail in the Toothbrush Cluster

That is the name for an unusual mix of objects so the result is a "cluster."

Astronomers have observed a "double strand' with a 'twist' coming from a 'junction' with radio radiation "revealing long filamentary structures."

These electric cosmology terms were a surprise in this AAS journal.

These details are observed in radio in a feature called a "toothbrush."

Clicking on the lower image reveals the named details.

Several segments attached to this structure are called "bristles."

In another Chandra image these bristles maintain their rough tubular shape for some distance.

Astronomers believe this structure with its "unusual shapes'" is from a possible merger. Details are missing.

The distracting blue cloud in the top image is supposedly dark matter where expected.

Below that image is a line in the text to click for a composite image.
This composite image has two very intense X-ray sources, below the "tooth brush." One must assume this pair along with the toothbrush are the 3 items in this cluster.

The story is about only the tooth brush feature so this pair is oddly ignored. Also strange is only the left one, not the right one, is seen in the top image.

Perhaps, these 2 actually plasmoids like in M87.

Excerpt:

The most prominent radio feature extends over more than six million light years, with three distinct components that resemble the brush and handle of a toothbrush. The handle is particularly enigmatic because, besides being large and very straight, it is off center from the axis of the cluster.
(Excerpt end)

Observation:

There is another story about this galaxy with the title: Deep VLA Observations of the Toothbrush Cluster

Excerpt:

Astronomers studying the Toothbrush Cluster with new radio observations combined with other wavelengths have been able to confirm the galaxy merger scenario and estimate the magnetic field strength in the shocks.
 There is more mass in this gas than there is in all the stars of a cluster's galaxies, and the gas can have a temperature of ten million kelvin or even higher.

As a result, the gas plays an important role in the cluster's evolution. The hot intracluster gas contains rapidly moving charged particles that radiate strongly at radio wavelengths, sometimes revealing long filamentary structures.
(Excerpt end)

Observation:
Cosmologists cannot explain this synchrotron radiation so they are left with impossible temperatures or an undefined mechanism with "shocks."

The observed "double strand with a twist" implies a pair of Birkelund filaments.
This structure is very long, at 6Mly.
Synchrotron radiation extending from radio to X-ray can be from fast electrons bending their path in a magnetic field even in intergalactic space.
Cosmologists have only one mechanism, an incredible, impossible temperature.

A published observation of a double strand and of long filamentary structures is notable. These are evidence of plasma.

The two unmentioned intense X-ray sources being out in space and not among stars suggest the two are plasmoids. They certainly do not look like a galaxy.

# 8 Neutron Star

Wikipedia excerpt:

A neutron star is the collapsed core of a massive supergiant star, which had a total mass of between 10 and 25 solar masses, possibly more if the star was especially metal-rich. Neutron stars are the smallest and densest stellar objects, excluding black holes and hypothetical white holes, quark stars, and strange stars. Neutron stars have a radius on the order of 10 kilometres (6.2 mi) and a mass of about 1.4 solar masses. They result from the supernova explosion of a massive star, combined with gravitational collapse, that compresses the core past white dwarf star density to that of atomic nuclei. The origins of the strong magnetic field are as yet unclear.

(Excerpt end)

Observation:
This entire explanation is invalid. There is no evidence many neutrons with no protons will not decay just like a solitary neutron decays in a few minutes. The description admits the magnetic field cannot be explained.


Thunderbolts Project has a YouTube video clearly explaining a neutron star, titled:

The Invention of the Neutron Star | Space News

Reference section at the end of this book has links.

The first book detailed the false claims of LIGO which never detected a non-existent gravitational wave, though LIGO made false claims of mergers with neutron stars.

Neutron stars do not exist.

# 9 Emission Nebula

An emission nebula has ions emitting radiation when they capture electrons, usually in ultraviolet but sometimes in a visible frequency.

Excerpt from Wikipedia:

An emission nebula is a nebula formed of ionized gases that emit light of various wavelengths. The most common source of ionization is high-energy ultraviolet photons emitted from a nearby hot star. Among the several different types of emission nebulae are H II regions, in which star formation is taking place and young, massive stars are the source of the ionizing photons; and planetary nebulae, in which a dying star has thrown off its outer layers, with the exposed hot core then ionizing them.
(Excerpt end)

# 10 Nursery Nebula

A Nursery nebula is assumed to have clouds of gas and dust called star formation regions.

Orion Nebula and the Eagle Nebula, aka the Pillars of Creation are the most famous.

## 10.1 Orion Nebula

Excerpt from Wikipedia for Orion Nebula:

The Orion Nebula is one of the most scrutinized and photographed objects in the night sky, and is among the most intensely studied celestial features. The nebula has revealed much about the process of
how stars and planetary systems are formed from collapsing clouds of gas and dust. Astronomers have directly observed protoplanetary disks, brown dwarfs, intense and turbulent motions of the gas, and the photo-ionizing effects of massive nearby stars in the nebula.

(Excerpt end)

## 10.2 Eagle Nebula

Chandra story is titled:

'X'-ploring the Eagle Nebula and 'Pillars of Creation

The image from Chandra has many X-ray point sources.

Excerpt:
The Chandra data reveal X-rays from hot outer atmospheres from stars. In this image, low, medium, and high-energy X-rays detected by Chandra have been colored red, green, and blue.

(Excerpt end)

Observations:

High energy X-rays come from a source of energetic synchrotron radiation, definitely not from heat.

There is a bright blue source near the top of the left pillar. The outer edge above this blue object looks rather like the solar corona, electric discharge activity with the plasma visible in diffused clouds or sheets.

The top of the central pillar has what looks like a coronal loop.
The top of the right pillar has a V with a stronger plasma discharge than along the edges below it.

Near the lower left region in the image are 2 blue objects, one above the other, with the lower one on the edge of the structure probably not by coincidence. The lower object might be the knot mentioned below.

Directly to the right is a luminous plasma filament.
Wikipedia has a topic for:
Eagle Nebula

Here is a much wider field showing the many bright stars behind these pillars.

References offers a link to only the high resolution image of the Eagle Nebula, from Wikipedia as a wide field.

In the wide field, there are quite a few "stars in a line" suggesting the electric cosmology formation method between a Birkelund current pair resulting in a string of stars, might be found in this image.

The background for these pillars implies the electrical discharges observed along the nebula edges are driven by the many bright stars behind these structures.

These structures in the nebula must have bonds to maintain their shape and edges suggesting plasma not neutral particles.

from Wikipedia:
the "Pillars of Creation" [is] a large region of star formation.

These columns – which resemble stalagmites protruding from the floor of a cavern – are composed of interstellar hydrogen gas and dust, which act as incubators for new stars.

Inside the columns and on their surface astronomers have found knots or globules of denser gas, called EGGs ("Evaporating Gaseous Globules"). Stars are being formed inside some of these EGGs.

(Excerpt end)

Observation:

Stars are composed of metallic hydrogen, the lattice of protons maintained by electrons.

There is no dust needed in making a star. The EGG sounds like denser hydrogen leading to metallic hydrogen with more plasma under compression.

If dust clouds are compressed, the result could be planets but not stars.

These "pillars of creation" might be creating stars along the edges where the electrical activity increases. The required compression probably will not occur within the pillars where the gas and dust mix is described as an incubator for a star's creation.

Dust is probably not actually needed in creating a star.

However the pillars suggest the electrical discharge activity along the edges of the dust structure is involved so the dust as plasma apparently facilitates in the electrical activity which is required for compression.

# 11 Planetary Nebula

A planetary nebula is the ejection of the outer layer of plasma from a star.

Excerpt from Wikipedia:

A planetary nebula is a type of emission nebula consisting of an expanding, glowing shell of ionized gas ejected from red giant stars late in their lives.

(Excerpt end)

Crab Nebula or Messier 1 was described as a supernova in section 6.1

## 11.1 Individual Planetary Nebulae

Plasma is not explicitly associated with a planetary nebula but this association is clear.

References has a link for any of interest.

Excerpt from Wikipedia (and most others below):

A planetary nebula, abbreviated as PN or plural PNe, is a type of emission nebula consisting of an expanding, glowing shell of ionized gas ejected from red giant stars late in their lives.

The Hubble Space Telescope showed that while many nebulae appear to have simple and regular structures when observed from the ground, the very high optical resolution achievable by telescopes above the Earth's atmosphere reveals extremely complex structures.
(Excerpt end)

Observation:
"Ionized gas" is plasma and plasma naturally forms filaments so the resulting structures from plasma behaviors are photogenic.

Images taken in ultraviolet or X-ray reveal the electrical activity in these shells of gas because synchrotron radiation spans wavelengths beyond the visible range.

Excerpt from Chandra page titled "A planetary Nebula Gallery" dated 10.10.12 :

twenty one planetary nebulas within about 5000 light years of the Earth have been observed. The paper also includes studies of fourteen other planetary nebulas, within the same distance range, that Chandra had already observed.

About half of the planetary nebulas in the study show X-ray point sources in the center, and all but one of these point sources show high energy X-rays that may be caused by a companion star, suggesting that a high frequency of central stars responsible for ejecting planetary nebulae have companions.

(Excerpt end)

Observation:
Another possible X-ray point source is a plasmoid rather than a binary somehow generating higher frequency radiation extending to X-ray and appearing as a single point source.

Following are some interesting planetary nebulae. Many have a similar pattern of two halves in the overall structure.

References has the link for any of interest.

M27 Dumbbell Nebula - has a rough rectangular shape with a filament dividing it in half.

Excerpt: "Similarly to the Helix Nebula and the Eskimo Nebula, the heads of the knots have bright cusps which are local photoionization fronts."

M57 Ring Nebula - at high resolution the ring has filaments across the middle, dividing the ring in half.

M76 Little Dumbbell Nebula - has a rough "figure 8" shape
M97 Owl Nebula - the eyes of the owl are split by the central bridge

NGC 6543 Cat's Eye Nebula - complicated arrangement of filaments as separate arcs

IC 3568 Lemon Slice Nebula - is a complete sphere inside another sphere with the inner showing many defined filaments while the outer sphere is diffused.

ngc 2392 Eskimo Nebula - complex inner structure (not symmetrical) while "[the] outer disk contains unusual, light-year-long filaments."

Necklace nebula - its ring has structure extending away the central star

NGC 7009 Saturn Nebula - several ellipses with active nodes at their ends.

Excerpt: "The nebula was originally a low-mass star that ejected its layers into space, forming the nebula."

NGC 7293 Helix Nebula - ring are "cometary" filaments extending from the central star

Excerpt:

Its main ring contains knots of nebulosity, which have now been detected in several nearby planetary nebulae, especially those with a molecular envelope like the Ring nebula and the Dumbbell Nebula. These knots are radially symmetric (from the [central star] and are described as "cometary", each centered on a core of neutral molecular gas and containing bright local photoionization fronts or cusps towards the central star and tails away from it.

(Excerpt end)

NGC 5189 Spiral Nebula - has a "figure 8" shape but some arcs are broken

Observation:

Offers only a mention of the "point symmetric knots" and "two dense low-ionization regions" but without further explanation.

Observation to the above collection:

Some of the explanations for these PNe structures are awkward.

Chandra has an image of 4 planetary nebulae in a page cited at the top, showing multiple wavelengths:

There is one star type often associated with planetary nebula: WR.

Excerpt from Wikipedia:

Wolf–Rayet stars, often abbreviated as WR stars, are a rare set of stars with unusual spectra showing prominent broad emission lines of ionised helium and highly ionised nitrogen or carbon. The spectra indicate very high surface enhancement of heavy elements, depletion of hydrogen, and strong stellar winds. Their surface temperatures range from 30,000 K to around 210,000 K, hotter than almost all other stars. They were previously called W-type stars referring to their spectral classification.

Classic (or Population I) Wolf–Rayet stars are evolved, massive stars that have completely lost their outer hydrogen and are fusing helium or heavier elements in the core. A subset of the population I WR stars show hydrogen lines in their spectra and are known as WNh stars; they are young extremely massive stars still fusing hydrogen at the core, with helium and nitrogen exposed at the surface by strong mixing and radiation-driven mass loss.

A separate group of stars with WR spectra are the central stars of planetary nebulae (CSPNe), post asymptotic giant branch stars that were similar to the Sun while on the main sequence, but have now ceased fusion and shed their atmospheres to reveal a bare carbon-oxygen core.

Only a minority of planetary nebulae have WR type central stars, but a considerable number of well-known planetary nebulae do have them.

(Excerpt end)

Observation:

The Wikipedia topic has a list of "List of Planetary Nebulae" Not all are included here.

The fusion model for stars (ignoring plasma behaviors) hinders how these planetary nebulae are explained, relying on "strong stellar winds" but these structures are not formed by a meteorological process.

Some WR stars even "ceased fusion" so only the "bare core" remains as the source of visible light. This is awkward.

This collection is interesting when knowing uncontrolled ejections and gravity alone cannot create such structures. The popular explanations are inadequate.

Plasma naturally forms filaments.

NGC 2452 is a 5-sided object with filaments and holes
Excerpt:

The blue haze across the frame is what remains of a star like our Sun after it has depleted all its fuel. When this happens, the core of the star becomes unstable and releases huge numbers of incredibly energetic particles that blow the star's atmosphere away into space. At the centre of this blue cloud lies what remains of the nebula's progenitor star. This cool, dim, and extremely dense star is actually a pulsating white dwarf, meaning that its brightness varies over time as gravity causes waves that pulse throughout the small star's body.

Observation:
A pulsating star with gravity causing waves affecting the brightness is an odd explanation.

NGC 2867 is a sphere around a star but no explanation is found.

NGC 1371 - 1372 is a double lobed object given two NGC numbers. NGC 5315 is an irregular cloud having a rough square hole around the central star.

NGC 7026 is a rough rectangular cloud with a rough square hole to the left of the central star.

NGC 1501 Oyster Nebula is a rough sphere having structure

NGC 6751 Glowing Eye Nebula is a inner and outer rings each having a complex filamentary structure

NGC 6369 Little Ghost Nebula has a wider inner ring with a diffused outer ring

MCn18 Engraved Hourglass Nebula has multiple non-concentric rings

Observation:
Synchrotron radiation implies a plasmoid not failed star.


Butterfly nebula

Part of the planetary nebula NGC 2899 has the rough shape of a butterfly. The oxygen ion emission line in UV enables its presentation with the false color of blue.
When protons capture electrons the result is either Lyman or Ballmer series emission lines, depending on the orbital the electron takes.
The red fringe can come from the hydrogen Balmer-alpha emission line which is truly in red.
Lyman-alpha emission line is in ultraviolet.
Ballmer-beta line is aqua while Balmer-gamma is blue.

Story with false colored image; link in References

Caption for the image:

This highly detailed image of the fantastic NGC 2899 planetary nebula was captured using the FORS instrument on ESO's Very Large Telescope in northern Chile. This object has never before been imaged in such striking detail, with even the faint outer edges of the planetary nebula glowing over the background stars.

Excerpt from the story:

This particular nebula is at a very high temperature, with the hot gas glowing to create the visual effect.

"NGC 2899's vast swathes of gas extend up to a maximum of two light-years from its center, glowing brightly in front of the stars of the Milky Way as the gas reaches temperatures upwards of 10,000 degrees [Celsius]," ESO scientists explained in a statement. "The high temperatures are due to the large amount of radiation from the nebula's parent star, which causes the hydrogen gas in the nebula to glow in a reddish halo around the oxygen gas, in blue."

Astronomers believe that the nebula developed its unusual shape because it has two central stars, which push out and illuminate gas in a symmetrical way. This type of nebula is called bipolar, and only around 10% to 20% of nebulae are of this type.

The image was captured using the FORS instrument on the VLT, standing for FOcal Reducer and low dispersion Spectrograph, which images in the visual and near-ultraviolet light wavelengths.

(Excerpt end)

Observation:

There is no glowing at ridiculous temperatures from distant stars.
The red on the fringes comes from protons capturing electrons. The UV from the oxygen is from their capture of electrons. Just because the atoms are ions does NOT mean these ions are at the ridiculous temperature.

There are no stars "which [can] push out and illuminate gas in a symmetrical way."
There is no wind in space.

The sharp edge on the nebula cannot be just gas and dust. Any edge cannot be anything other than a liquid or solid. Loose particles cannot form and maintain an edge until bonded.

A likely possibility is the internal edges of the nebular are condensed matter, or metallic hydrogen (just protons and electrons) which can be liquid when not compressed to a solid.

The diffused glow near the edge suggests electrical activity like in the solar corona.

The image from APOD reveals this entire nebula consists of plasma filaments.

Here the "butterfly" is slightly below and right of center and is without the false blue color.

Unlike the other image, APOD shows red and other colors everywhere else but in the butterfly.

Artists make these celestial objects more interesting.

Its caption mentions the ridiculous temperature:

Can stars, like caterpillars, transform themselves into butterflies? No, but in the case of the Butterfly Nebula -- it sure looks like it. Though its wingspan covers over 3 light-years and its estimated surface temperature exceeds 200,000 degrees C, the dying central star of NGC 6302, the featured planetary nebula, has become exceptionally hot, shining brightly in visible and ultraviolet light but hidden from direct view by a dense torus of dust. This sharp close-up was recorded by the Hubble Space Telescope and is reprocessed here to show off the remarkable details of the complex planetary nebula, highlighting in particular light emitted by iron, shown in red. NGC 6302 lies about 4,000 light-years away in the constellation of the Scorpion. Planetary nebulas evolve from outer atmospheres of stars like our Sun, but usually fade in about 20,000 years. (Caption end)

Observation:

The iron emission line is not red so this is a false color from what an eye will see.

"Planetary nebulas [do not] evolve from outer atmospheres of stars like our Sun."

The complex filamentary structure in the nebula was definitely not a slow evolution from just any star.

Another image from NASA offers different detail:

Excerpt:

NGC 6302 is no exception. With an estimated surface temperature of about 250,000 degrees C, the dying central star of this particular planetary nebula has become exceptionally hot, shining brightly in ultraviolet light but hidden from direct view by a dense torus of dust. This sharp close-up of the dying star's nebula was recorded by the Hubble Space Telescope and is presented here in reprocessed colors. Cutting across a bright cavity of ionized gas, the dust torus surrounding the central star is near the center of this view, almost edge-on to the line-of-sight. Molecular hydrogen has been detected in the hot star's dusty cosmic shroud. NGC 6302 lies about 4,000 light-years away.

(Excerpt end)

Observation:

The hottest star type is O-type with a surface temperature of 30,000 to 60,000 K.

The central star's temperature is literally "off the chart" but no spectrum is provided to check how the extreme temperature was measured. It is odd this "dying star" overheats without exploding, or failing in some manner.

Everything in the butterfly nebula is plasma, with ions and protons capturing electrons to emit radiation at particular wave lengths.

# 12 Reflection Nebula

Excerpt from Wikipedia:

In astronomy, reflection nebulae are clouds of interstellar dust which might reflect the light of a nearby star or stars. The energy from the nearby stars is insufficient to ionize the gas of the nebula to create an emission nebula, but is enough to give sufficient scattering to make the dust visible. Thus, the frequency spectrum shown by reflection nebulae is similar to that of the illuminating stars. Among the microscopic particles responsible for the scattering are carbon compounds (e. g. diamond dust) and compounds of other elements such as iron and nickel. The latter two are often aligned with the galactic magnetic field and cause the scattered light to be slightly polarized.

(Excerpt end)

## 12.1 IC 2118 is a well known reflection nebula.

Excerpt from Wikipedia:

IC 2118 (also known as Witch Head Nebula due to its shape) is an extremely faint reflection nebula believed to be an ancient supernova remnant or gas cloud illuminated by nearby supergiant star Rigel in the constellation of Orion. It lies in the Orion constellation, about 900 light-years from Earth. The nature of the dust particles, reflecting blue light better than red, is a factor in giving the Witch Head its blue color. Radio observations show substantial carbon monoxide emission throughout parts of IC 2118.

(Excerpt end)

Observation:

If this nebula is really a supernova remnant than it began as plasma, the ejected outer layer of the star which erupted.

Astronomy Picture of the day on 2009 December 29 was titled:

Rigel and the Witch Head Nebula

Excerpt:

Explanation: Double, double toil and trouble; Fire burn, and cauldron bubble -- maybe Macbeth should have consulted the Witch Head Nebula. This suggestively shaped reflection nebula on the lower left is associated with the bright star Rigel, to its right, in the constellation Orion. More formally known as IC 2118, the Witch Head Nebula glows primarily by light reflected from Rigel. Fine dust in the nebula reflects the light. Pictured above, the blue color of the Witch Head Nebula and of the dust surrounding Rigel is caused not only by Rigel's blue color but because the dust grains reflect blue light more efficiently than red. The same physical process causes Earth's daytime sky to appear blue, although the scatters in Earth's atmosphere are molecules of nitrogen and oxygen. Rigel, the Witch Head Nebula, and gas and dust that surrounds them lie about 800 light-years away.

(Excerpt end)

Observation:

The APOD image shows a great distance between the nebula and any bright stars. One could suggest this brightness cannot be attributed to only reflection because scattering cannot be perfect without losses. However with no spectrum data, it is impossible to determine if some amount of the brightness is from emission lines.

The Wikipedia entry mentions radio emissions which must come from synchrotron radiation, which requires plasma in motion.

## 12.2 Trifid Nebula

Trifid Nebula has 3 lobes, with one as a reflection nebula, another as an emission nebula, with a third lobe as a "dark" nebula.

Excerpt from Wikipedia:

The Trifid Nebula (catalogued as Messier 20 or M20 and as NGC 6514) is an H II region located in Sagittarius. Its name means 'divided into three lobes'. The object is an unusual combination of an open cluster of stars; an emission nebula (the lower, red portion), a reflection nebula (the upper, blue portion) and a dark nebula (the apparent 'gaps' within the emission nebula that cause the trifurcated appearance; these are also designated Barnard 85). Viewed through a small telescope, the Trifid Nebula is a bright and peculiar object, and is thus a perennial favorite of amateur astronomers.

The Trifid Nebula was the subject of an investigation by astronomers using the Hubble Space Telescope in 1997, using filters that isolate emission
from hydrogen atoms, ionized sulfur atoms, and doubly ionized oxygen atoms. The images were combined into a false-color composite picture to suggest how the nebula might look to the eye.
The close-up images show a dense cloud of dust and gas, which is a stellar nursery full of embryonic stars. This cloud is about 8 ly away from the nebula's central star. A stellar jet protrudes from the head of the cloud and is
about 0.75 ly long. The jet's source is a young stellar object deep within the cloud. Jets are the exhaust gasses of star formation and radiation from the nebula's central star makes the jet glow.

The images also showed a finger-like stalk to the right of the jet. It points from the head of the dense cloud directly toward the star that powers the Trifid nebula. This stalk is a prominent example of evaporating gaseous globules, or 'EGGs'. The stalk has survived because its tip is a knot of gas that is dense enough to resist being eaten away by the powerful radiation from the star.

(Excerpt end)

Observation:

A jet is a plasma filament, not "exhaust gasses." "powerful radiation" can create ions which lead to plasma behaviors like creating filaments. If the jet glows, that is the result of ions capturing electrons causing emission lines. Some are visible like in the Earth's aurora.

# 13 Quasar

Excerpt from Wikipedia:

A quasar (also known as a quasi-stellar object abbreviated QSO) is an extremely luminous active galactic nucleus (AGN), in which a supermassive black hole with mass ranging from millions to billions of times the mass of the Sun is surrounded by a gaseous accretion disk. As gas in the disk falls towards the black hole, energy is released in the form of electromagnetic radiation, which can be observed across the electromagnetic spectrum. The power radiated by quasars is enormous: the most powerful quasars have luminosities thousands of times greater than a galaxy such as the Milky Way.
(Excerpt end)

Note author's quasar model was in the second book, including a detailed analysis of the spectrograms of quasars in Halton Arp's book Seeing Red.

BeppoSAX did a study of quasars and BL Lac objects, titled:

BeppoSAX observations of synchrotron X-ray emission from radio quasars

The conclusion is both have an active galactic nucleus emitting synchrotron radiation extending to X-ray frequencies.

References at the end of this book includes the author's quasar hypothesis which was the preliminary basis for that section of the second book. The book has the final description.

Here is a short story of quasars. Some of this was described in earlier sections, and in the second book in the series, about the transition in cosmology.

Active spiral galaxies are observed to eject plasmoids.

Seyfert galaxies are active spiral galaxies which are in a class called LINER because their nucleus has many metallic ions.

Halton Arp observed Seyfert galaxies appeared by association to eject quasars, often in a pair in opposing directions along the galaxy axis.

A quasar is a plasmoid shrouded in a cloud of metallic atoms resulting in dimming in optical, but the plasmoid is often intense from X-ray to radio.

The quasar exhibits two distinct red shifts, with one from the metallic ions moving toward the plasmoid. This red shift decreases over time as the cloud of ions disperses. Arp observed this decrease seemed to be in increments rather than linear or random.

A second, higher red shift, results from high velocity protons approaching the plasmoid and capturing an electron, generating the hydrogen emission line with a red shift. Quasar red shifts of $z > 7$ have been measured. This velocity is that of the proton, not the quasar.

# 14 Black Hole

This fictitious entity was covered in both books.

A black hole has a simple purpose in popular cosmology.

Plasma and electromagnetic forces are ignored so synchrotron radiation is also ignored. As a result there is no available source for X-rays.

A black hole enables an impossible source of X-rays as a theoretical result of an extremely hot object. The solution was using a black hole which could supposedly hold an impossible mass within a point, or a sphere of zero diameter.

Excerpt from Wikipedia:

A black hole is a region of spacetime where gravity is so strong that nothing—no particles or even electromagnetic radiation such as light—can escape from it. The theory of general relativity predicts that a sufficiently compact mass can deform spacetime to form a black hole. The boundary of the region from which no escape is possible is called the event horizon. Although the event horizon has an enormous effect on the fate and circumstances of an object crossing it, according to general relativity it has no locally detectable features. In many ways, a black hole acts like an ideal black body, as it reflects no light. Moreover, quantum field theory in curved spacetime predicts that event horizons emit Hawking radiation, with the same spectrum as a black body of a temperature inversely proportional to its mass.

This temperature is on the order of billionths of a kelvin for black holes of stellar mass, making it essentially impossible to observe.
(Excerpt end)

Observation:

All is "according to relativity" which involves only the moving observer's reference frame. No other observer observes curvature, so no other observer, like one on or near Earth can observe the curvature at a black hole.

The impossible claims of a black hole were detailed in the author's first book. Many others have their own way to justify the inevitable conclusion there are no black holes.

Even Albert Einstein knew a black hole is impossible, with his famous quote:

"Black holes are where God divided by zero."

A black hole is fictional.

As noted in the section about quasars, a study looked for thermal radiation from a supposed black hole in several quasars and BL Lac objects which are similar to quasars.

All had synchrotron radiation indicating a plasmoid and no black hole with its impossible accretion disk.

The infamous fabricated image in April 2020 of an object in M87 galaxy was not a black hole. It was a plasmoid, covered in a previous section.

No black hole has ever been detected because they don't exist.

The first book detailed the false claims of LIGO which never detected a non-existent gravitational wave, though LIGO made false claims of mergers with black holes. LIGO never detected an event with a black hole or neutron star.


A story in April 2020 was titled:
Astronomers saw a star dancing around a black hole. And it proves Einstein's theory was right

Observation:

For many years astronomers have been seeking a star orbiting around the claimed black hole at the center of our Milky Way. If its orbit is a valid Kepler ellipse then they seek to apply Kepler's 3rd law to calculate the MBH mass. This was done with S1 years ago to derive the silly mass of billions of solar masses. Unfortunately someone noticed S1 was not a valid ellipse; instead of the claimed period of 16 years, its claimed radius required many thousands of years. The silly mass persists but astronomers seek another star to try again for this mass calculation.

Obviously S2 orbit is invalid as well because it is not closed as stated in the article.

Because astronomers have been watching for a few decades, so perhaps they gave up and came up with this "rosette" orbit to bring up Einstein - which is obviously effective to get your story published.

From Wikipedia, not the story:

S2, also known as S0–2, is a star that is located close to the radio source Sagittarius A*, orbiting it with an orbital period of 16.0518 years, a semi-major axis of about 970 au.

(Excerpt end)

Observation:

S2 = 970 au, 16.052 y

Jupiter= 5.203au,  11.86 y
Saturn= 9.54au, 29.46y

With Kepler's 3rd law:

S2 = 970 au takes  about 30,200yr

S2= 16.0518 yr is for about 6.5au

the observed orbit parameters for S2 are not a valid Kepler ellipse.

These news stories never reveal the importance of this search for a star like S2 and the requirement for a valid ellipse.

With a valid ellipse, astronomers can use Newton's change to Kepler's 3rd law to calculate the number of solar masses at the core with a known mass in an orbiting star. The mass for S2 will be a guess. With electromagnetic forces in play this calculation might be debated for many reasons.

Currently, every black hole claimed to be in every galaxy is assigned a value of solar masses loosely based on the estimated number of stars in that galaxy. That assumption has no basis.

If the Milky Way core, which is claimed to have a SMBH for the roughly 300 billion stars in the galaxy, is monitored to seek an orbiting star. When finally found, this will be a significant milestone for black hole advocates.

Until that star is found, there is no black hole in any galaxy with a justification for its claimed number of solar masses.

This is a basic problem for cosmologists of no evidence or no justification for any claims about a SMBH in every galaxy. Cosmologists hope this finding and calculation can salvage all their claims spanning many years. Unfortunately for them, an orbit with a larger radius for its measurement requires a longer period like in thousands of years (like S2 needs).

One might expect rules to be broken about the orbit parameters to achieve the crucial calculation.

Every scientific claim requires evidence.

Currently, there is no evidence for any value of a black hole's mass

# 15 Galaxy

For those who dud not read the first 2 books, a spiral galaxy rotates by the galactic magnetic field and there is no dark matter.

The Journal Progress in Physics in April 2018 had the paper by Donald Scott titled:

Birkelund Currents and Dark Matter

Pdf link in References

Figure 6 in the paper shows how well the predictions for a spiral galaxy rotation closely match observations.

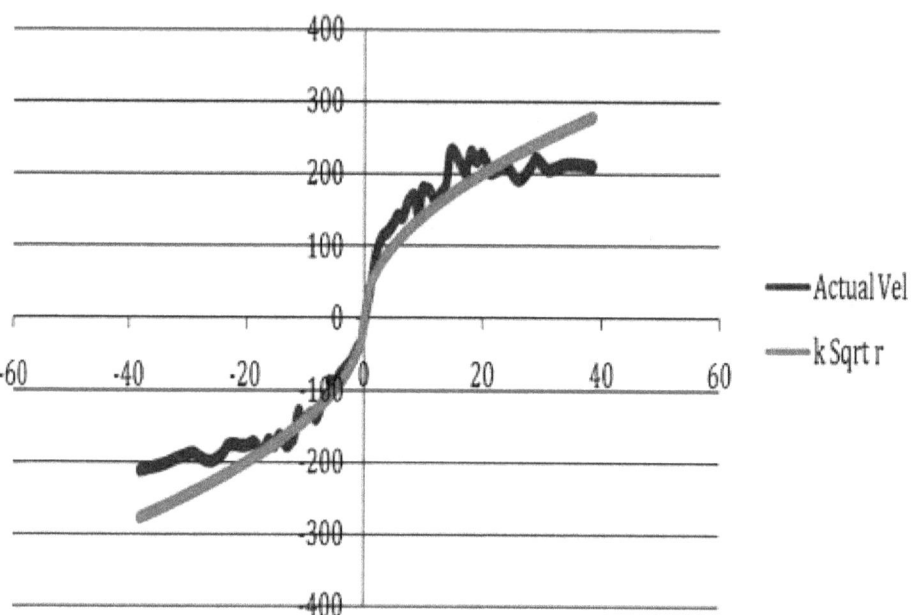

Fig. 6: Comparison of the example galaxy's measured velocity profile with the Bessel function model's Sqrt r profile.

The paper provides details.

For those needing more references about a spiral galaxy rotation curve, scientists in Spain published the paper titled:

MAGNETIC FIELDS AND THE OUTER ROTATION CURVE OF M31

Excerpt:

It is certainly a challenge as the standard dark matter halo models, in particular the universal NFW profiles, do not account for this dynamical unexpected behavior.
Our conclusion is that a significant improvement of the fit in the outer part is obtained when magnetic effects are considered. The best-fit solution requires an amplitude of ~4 µG with a weak radial dependence between 10 and 38 kpc.

(Excerpt end)


Observation:
Even the "dark matter halo" fails to account for "this unexpected behavior" but the "magnetic effect" is the "best-fit solution."

## 15.1 Dwarf Galaxy or Globular Cluster

That was the headline of a news story.

A new object was found and there is a debate whether it is a globular cluster or a dwarf galaxy; in either case it is hiding behind the obstruction of the Milky Way central bulge.

The debate reveals our assumptions for these objects.

Astronomers have access to new data available with a variety of recent instruments as part of sky surveys gathering details for many objects with each scan. That recent data enabled this discovery.

In this story FSR 1758 is either a dwarf galaxy or a globular cluster. The researchers have named it Scorpius dwarf galaxy.

On a first impression those two objects should be easily distinguished. They are not because FSR 1758 is so obscured it is difficult to determine the correct locations of its stars in 3-D; this is required to analyze stars in the object rather than just in the line of sight or just seem to be around it.

Excerpt:

Spatial distribution of stars that have common proper motion, suggesting that they belong to FSR 2758. Beyond the visible custer of stars at the center of the group, FSR 1758 appears to have a possible larger extended structure, suggesting it may be the nucleus of a dwarf galaxy.
FSR 1758 has a number of intriguing properties. If it's a globular cluster, it's one of the largest around our galaxy — and the part that we see is probably just the metaphorical tip of the iceberg, as much of its population is likely hidden by contamination and reddening due to its location in the galactic bulge. Furthermore, FSR 1758's properties don't fit known relationships for globular clusters, such as the correlation between size and metallicity.

Lastly, the authors find additional asymmetrically distributed stars further out in the field with motions and colors indicating that they also belong to FSR 1758. These suggest that the cluster may be more extended than originally thought and might have tidal tails. These signs support a picture in which FSR 1758 is the nucleus of a dwarf galaxy — which the authors tentatively name the Scorpius dwarf galaxy.
Though we still don't have a definitive answer about FSR 1758's nature, we can hope that future spectral data for its stars will settle the debate.
(Excerpt end)

Observation:

Metallicity is critical to every analysis; it is the basis for our understanding stars. The story explicitly mentions this problem. "FSR 1758's properties don't fit known relationships for globular clusters, such as the correlation between size and metallicity."

From this statement, one could wonder whether the conclusion this object is a dwarf galaxy was based on the finding it is not compatible with our "understanding" of a globular cluster(GC).

The story mentions FSR 1758 could be one of the largest GC near the Milky Way.

Excerpt from Wikipedia:

Omega Centauri (or NGC 5139) is a globular cluster in the constellation of Centaurus located at a distance of 15,800 light-years, it is the largest globular cluster in the Milky Way at a diameter of roughly 150 light-years. It is estimated to contain approximately 10 million stars and a total mass equivalent to 4 million solar masses, making it the most massive globular cluster of the Milky Way.
Omega Centauri is so distinctive from the other galactic globular clusters that it is thought to have an alternative origin as the core remnant of a disrupted dwarf galaxy.

The stars in the core of Omega Centauri are so crowded that they are estimated to average only 0.1 light years away from each other. The internal dynamics have been analyzed using measurements of the radial velocities of 469 stars. The members of this cluster are orbiting the center of mass with a peak velocity dispersion of 7.9 km/s. The mass distribution inferred from the kinematics is slightly more extended than, though not strongly inconsistent with, the luminosity distribution.
(Excerpt end)

Observation:
One must note Omega Centauri has uncertainty about its origin as a GC. It is also distinctive among the GC implying a lack of consistent observations among them. One could wonder which is the prototype for other GC to be compared.

The density of stars in the core is both surprising and impressive.

Continuing Omega Centauri excerpt with part 2:

Using instruments at the Gemini Observatory to measure the speed of stars swirling in the cluster's core, E. Noyola and colleagues found that stars closer to the core are moving faster than stars farther away. This measurement was interpreted to mean that unseen matter at the core is interacting gravitationally with nearby stars. By comparing these results with standard models, the astronomers concluded that the most likely cause was the gravitational pull of a dense, massive object such as a black hole. They calculated the object's mass at 40,000 solar masses.
(Excerpt end 2)

Observation:

One could expect this discovery will be compared to Omega Centauri. That new GC could be unique or its conclusions are wrong.

The reference to "standard models" is problematic. A GC is not understood with a gravity cosmology which ignores electromagnetic forces.

Fictitious black holes should never be mentioned with GC stars.

One can only hope the status of FSR1758 will be done objectively, not distorted by assumptions which could be incorrect.

They claim future data will resolve the debate but historically many assumptions are rarely questioned.

An objective analysis is probably unlikely so the Milky Way probably gained a new dwarf galaxy.

## 15.2 NGC 1156

NGC 1156 is an interesting dwarf galaxy.

The story headline:
Hubble Views Lonely Dwarf Galaxy NGC 1156

Excerpt:

It is classified as a dwarf irregular galaxy, meaning that it lacks a clear spiral or rounded shape, as other galaxies have, and is on the smaller side, albeit with a relatively large central region that is more densely packed with stars. Some pockets of gas within NGC 1156 rotate in the opposite direction to the rest of the galaxy, suggesting that there has been a close encounter with another galaxy in the past.
(Excerpt end)

Observation:

NGC 1156 has interesting attributes unlike most spiral galaxies.
Among them:
a) no black hole,
b) no dark matter,
c) inconsistent rotation of the irregular shape.

Spiral galaxies are expected to rotate like our solar system where each object rotates about the barycenter, or the system's center of gravity.

Nearly every galaxy has an X-ray source at its core.
MC explains that source with a black hole with an extremely hot accretion disk. This black hole is always sized to be greater than the number of stars in the system to balance this assumed barycenter behavior.
MC requires black matter for every galaxy having rotation but the stars do not rotate about a barcenter so dark matter is the excuse for the wrong model for a spiral galaxy.

This particular galaxy has no X-ray source identified to justify (a).

The galaxy has an unusual rotation so there is no justification for (b).

Instead the anomalous rotation (c) is explained away by an unverifiable claim of an encounter with an unidentified galaxy with no details of this encounter, like the respective masses and velocities, trajectories, the distance, and time of this encounter.

The electric universe cosmology can offer believable explanations instead of this evasion.

Because there is an observed rotation there must be an axial electric current generating a galactic magnetic field exerting the Lorentz force for rotation.
There must be a Birkelund current in dark mode along this galaxy's axis.

There must be a Z-pinch at the core of this dwarf galaxy.

The Z-pinch is a source of synchrotron radiation. Its highest frequency is determined by the intensity of the electric current.
The story offers no spectrum details for this core.
From the description the core is obscured by the dense stellar activity.

Perhaps the electric current to this galaxy is inadequate for X-rays.

Donald Scott has a pod cast explaining how a Birkelund current filament with its multiple concentric current paths can result in counter rotating disks in a spiral galaxy.

NGC 1156 is another example of that behavior.

Because the galaxy rotation is present but not well defined, one explanation is a weaker magnetic field from the Birkelund current filament weaker than typical for a definite spiral galaxy appearance.

The Hubble image shows noticeable stellar activity so the galaxy must have enough current to sustain what is observed.

The story's mention of a "characteristic pink hue" is odd because the Lyman-alpha emission line of hydrogen is in ultraviolet.
However, the Ballmer-alpha line is red from a lower energy state for the electron.

Ultraviolet frequency will affect neutral hydrogen atoms, like in gas clouds near the core. Stars like our Sun radiate in ultraviolet. The story claims "newborn stars" are doing this but actually most stars (i.e., similar to the Sun or more energetic) are probably capable.

This particular color must be from the filters mentioned in the caption. That means some portion of this image has false colors. The caption makes this clear but not the story.

From the above observations NGC 1156, called a dwarf irregular galaxy, is just an under-powered spiral galaxy.

## 15.3 Magellanic Clouds

The two Magellanic Clouds are galaxies close to our Milky Way but their Southern position inhibits viewing by Northern telescopes.

Both galaxies are classified as SB(s)m, or a "modified" barred spiral galaxy.
From Wikipedia:

Announced in 2006, measurements with the Hubble Space Telescope suggest the Large and Small Magellanic Clouds may be moving too fast to be orbiting the Milky Way.
 In 2014, measurements from the Hubble Space Telescope made it possible to determine that the LMC has a rotation period of 250 million years.
(Excerpt end)

From a space.com story about LMC:

By pointing NASA's Hubble Space Telescope toward the two clouds, scientists began to catch a glimpse of the objects' histories. "Hubble's biggest contribution is enabling us to clock how fast the Magellanic clouds are moving," said Gurtina Besla, a researcher at the University of Arizona who studies dwarf galaxies. In 2007, Besla overturned conventional wisdom when she suggested that the LMC and SMC were making their first orbit of our galaxy.
"They're moving too fast to have been long-term companions of the Milky Way," Besla said.

She used data from the European Space Agency's Gaia spacecraft to clock smaller, satellite galaxies orbiting the LMC, as well. And, understanding how these galaxies move has helped researchers better calculate the mass of the LMC. Current estimates put the LMC at about 100 billion times as massive as Earth's sun, or a quarter the mass of the Milky Way. Besla said this size means the LMC is about 10 times heavier than previously calculated.


NASA has a database in its MAGCMXBCAT 'catalog of high-mass X-ray binaries (HMXBs) in the Small and Large Magellanic Clouds (SMC and LMC). The aim of this catalog is to provide easy access to the basic information on the X-ray sources and their counterparts in other wavelength ranges (UV, optical, IR, radio). Most of the sources have been identified as Be/X-ray binaries. Some sources, however, are only tentatively identified as HMXBs on the basis of a transient character and/or a hard X-ray spectrum. Further identification in other wavelength bands is needed to finally determine the nature of these sources. In cases where there is some doubt about the high-mass nature of the X-ray binary this is mentioned.
(Excerpt end)

Observation:
Apparently every X-ray source in the Magellanic Clouds is a "high-mass X-ray Binary."

Excerpt:
Be/X-ray binaries (BeXRBs) are a class of high-mass X-ray binaries that consist of a Be star and a neutron star. The neutron star is usually in a wide highly elliptical orbit around the Be star. The Be stellar wind forms a disk confined to a plane often different from the orbital plane of the neutron star. When the neutron star passes through the Be disk, it accretes a large mass of gas in a short time. As the gas falls onto the neutron star, a bright flare in hard X-rays is seen.

Observation:

Cosmologists have advanced to this mechanism: a Be star provides the disk for the neutron star to move through. When this gas accelerates it flares in X-ray. One could assume this accretion and flaring continues for millions of years.

Astronomers are always seeking the amount of mass in everything. The LMC just gained mass (10x). There are no defined criteria for defining the masses for each one of the binary in the BeXRB.

Astronomers follow a simple rule: every X-ray source is a black hole.

Their current accepted distances come from Cepheids not the red shift. Excerpt from Wikipedia:

Using this period-luminosity relation, in 1913 the distance to the SMC was first estimated by Ejnar Hertzsprung. First he measured thirteen nearby cepheid variables to find the absolute magnitude of a variable with a period of one day. By comparing this to the periodicity of the variables as measured by Leavitt, he was able to estimate a distance of 10,000 parsecs (30,000 light years) between the Sun and the SMC. This later proved to be a gross underestimate of the true distance, but it did demonstrate the potential usefulness of this technique.
(Excerpt end)

Observation:

When Hubble telescope provides resolution to individual stars the astronomers can ignore these red shifts. When an astronomer says "too fast" but the basis for that conclusion is not clear.

Astronomers are still learning about motion of the Magellanic Clouds.

30 Doradus is a well known nebula in the LMC. It is also known as Tarantula Nebula.

From NASA, a story:

30 Doradus - The Growing Tarantula Within

Excerpt:

The star-forming region, 30 Doradus, is one of the largest located close to the Milky Way and is found in the neighboring galaxy, Large Magellanic Cloud.

About 2,400 massive stars in the center of 30 Doradus, also known as the Tarantula Nebula, are producing intense radiation and powerful winds as they blow off material.

Multimillion-degree gas detected in X-rays by the Chandra X-ray Observatory comes from shock fronts - similar to sonic booms -formed by these stellar winds and by supernova explosions. This hot gas carves out gigantic bubbles in the surrounding cooler gas and dust.
(Excerpt end)

Observation:

X-rays are from a source of synchrotron radiation. All the rest with impossible temperatures and "shock" and "booms" is just nonsense,
The proliferation of impossible black holes (in an ever expanding variety) can be embarrassing.

However, winds, shock fronts, sonic booms, hot gas carving bubbles, and multimillion-degree gas emitting X-rays are not believable either.


A 2010 study of M31 concluded M31 has no dark matter, but instead the galactic magnetic field fits the rotation curve. This study was mentioned above, in the section about M31.

Cosmologists should have reacted to that finding, but did not, when reading current spiral galaxy descriptions.

## 15.1 M31 Andromeda Galaxy

Vera Rubin in the 1930's concluded M31 had dark matter to cause the observed rotation curve because all the visible mass could not account for the stars in their non-Keplerian orbits, meaning the stars in the galaxy disk are not in orbits like planets around the Sun. This assumption was invalid so dark matter resulted.

.

Excerpt from Wikipedia about M31:

The virial mass of the Andromeda Galaxy is of the same order of magnitude as that of the Milky Way, at 1 trillion solar masses. The mass of either galaxy is difficult to estimate with any accuracy, but it was long thought that the Andromeda Galaxy is more massive than the Milky Way by a margin of some 25% to 50%. This has been called into question by a 2018 study that cited a lower estimate on the mass of the Andromeda Galaxy, combined with preliminary reports on a 2019 study estimating a higher mass of the Milky Way.

(Excerpt end)

Observation:

Dark matter was proposed when lacking certainty of the galaxy's mass. Even by 2018, the mass estimate has a large margin of error.

The author's first book noted these studies and others to conclude there is no dark matter. Cosmology must drop that mistake.

Chandra made observations of the M31 core in a story titled:

M31: The Heat Is On in Andromeda's Center

Excerpt:

Analysis of the X-ray data shows that the point sources are associated with binary star systems that contain a neutron star or black hole that is pulling matter away from a normal star. As the matter falls toward the neutron star or black hole, it is heated by frictional forces to tens of millions of degrees, and produces X-rays.

The diffuse X-ray cloud is due to gas that has accumulated in the central region and been heated to millions of degrees, probably by shock waves from supernova explosions. The energy input from the supernovas could also be driving gas out of the central region. This process may affect both the shape and evolution of the galaxy by depleting the raw material for the formation of new stars and preventing more gas from accumulating there.

(Excerpt end)

Observation:

X-rays are always from synchrotron radiation and never from thermal radiation which never exceeds ultraviolet.

The core of Andromeda galaxy is like the core of the Milky Way with a pair of X-ray "chimneys" from the pair of axial Birkelund current filaments which bend and then split into individual paths for each spiral arm.

M31 is like the Milky Way in many observations including a barred spiral configuration.

## 15.4 IC342

A study revealed the importance of the magnetic fields in the spiral arms

The story:

Twisted magnetic field in galaxy IC 342

An interesting conclusion after a study of IC342, a large obscured, nearby spiral galaxy:

Excerpt:

Magnetic fields exist everywhere in the Universe, but what role do they play in the evolution of cosmic objects? In their detailed data of the nearby galaxy IC 342 from observations with two of the world's largest radio telescopes, Astronomers at the Max Planck Institute for Radio Astronomy (MPIfR) have discovered a magnetic field aligned along the optical spiral arms. "Our observations can help to discover answers to the question of how galaxies evolve and develop further," says project leader Rainer Beck.
"Spiral arms can hardly be formed by gravitational forces alone," continues Rainer Beck. "This new IC 342 image indicates that magnetic fields also play an important role in forming spiral arms."

(Excerpt end)

Excerpt from NOAO:

Spiral Galaxy IC342 is located roughly 11 million light-years from Earth. Its face-on appearance in the sky—as opposed to our tilted and edge-on views of many other nearby galaxies, such as the large spiral galaxy Andromeda (M31)—makes IC342 a prime target for studies of star formation and astrochemistry.

(Excerpt end)

Excerpt from another source:

[This] galaxy is obscured by the dust in our galaxy by nearly three full magnitudes. If it weren't for this dimming effect, IC342 would be one of the finest and brightest face-on spirals in the sky and easily visible to the unaided eye in dark sky. That would make it by far the most distant object you could see with the unaided eye.

(Excerpt end)

Observation:
The conclusion about the important role of magnetic fields presents a problem for cosmologists because those fields are usually not even looked for.

Currently the rotation curve of a spiral galaxy is assumed to be driven by the mass distribution in the disk.
This scientist claims gravity alone does not drive the spiral arm formation.

This is an important conclusion because when the mass distribution does not result in the correct rotation curve, dark matter is proposed to "fix" the "required" mass distribution.

Studies like this can lead to the removal of dark matter.

## 15.3 Galaxy core with jets

The story titled:

Chandra detection of a circumnuclear torus

Excerpt:
The combination of Chandra and ALMA observations found a torus emitting jets at the core of the active spiral galaxy NGC 5643.

From Wikipedia: "NGC 5643 has an active galactic nucleus and is a type II Seyfert galaxy."

The jets could be a surprise for a spiral galaxy.

M87 is an elliptical galaxy which famously had its torus or plasmoid imaged in April. M87 does not rotate like a spiral galaxy. The stars in an elliptical move radially, not in rotation about the sphere's center.

Our Milky Way is a spiral galaxy w/ith several rotating arms.

Donald Scott has explained how the Birkelund filament pair as the electric current along the galactic axis generates the galactic magnetic field for the observed rotation curve of the stars in the galactic disk.

At the Milky Way core is a Z-pinch of the filaments as the tiny source of energetic synchrotron radiation, extending to X-rays.

It is not generally known if jets have been observed at the Milky Way core.

## 15.4 Black Hole Jets

Scientific American had a story on 9/25/2013 titled "Milky Way's Black Hole Is Shooting Particle Jets"

Excerpt:

The Milky Way's giant black hole, called Sagittarius A* has long been theorized to have jets, but evidence was inconclusive. Now researchers have combined x-ray photographs of the galaxy's center from NASA's Chandra space telescope with radio data from the Very Large Array (VLA) observatory in New Mexico to offer the best support yet for the idea of jets from Sagittarius A*. The x-ray photos show a wispy bright line of gas that is emitting x-ray light to one side of the black hole—perhaps indicating the jet itself—and the radio observations highlight a wall of gas that scientists think is a shock front created where the jet is slamming into a cloud, snow-plowing the gas into a clump.

Jets arise because the black hole is spinning. As matter falls into the black hole, the matter's magnetic field gets twisted and amplified by the black hole's spin, and this pumped-up magnetic field launches material outward in the form of jets. If the signals from Chandra and the VLA really are a jet, its direction would reveal the spin axis of the Sagittarius A* black hole. "Lo and behold, the spin axis appears to be the same as the galaxy," Morris says. "That's so satisfying, because that's what you would expect if the black hole has never undergone a major disturbance."

Some previous studies, however, suggested the jets pointed in a different direction. The new study is well thought out and "will be a benchmark against future claims of jet orientations," says astronomer Heino Falcke of Radboud University Nijmegen in the Netherlands, who has also argued for a jet at the galactic center, and recently found that theoretical models favor one.

(Excerpt end)

Observation:

Of course the story is not plausible, with a spinning black hole, a "wall of gas is a shock front" then "snow-plowing the gas into a clump" - "because [about the spin axis] that's what you would expect"

The story concludes with uncertainty whether there are truly jets. Since NGC5643 has them so maybe we will find them here too.

## 15.5 Magellanic Clouds

The two Magellanic Clouds are galaxies close to our Milky Way but their Southern position inhibits viewing by Northern telescopes.
Excerpt from Wikipedia:

Announced in 2006, measurements with the Hubble Space Telescope suggest the Large and Small Magellanic Clouds may be moving too fast to be orbiting the Milky Way.
In 2014, measurements from the Hubble Space Telescope made it possible to determine that the LMC has a rotation period of 250 million years.

(Excerpt end)

Observation:

Such a period requires many life times to confirm.

## 15.5.1 LMC

A story about the motion of LMC, titled:

Large Magellanic Cloud: Nearby Satellite Dwarf Galaxy

Excerpt:

By pointing NASA's Hubble Space Telescope toward the two clouds, scientists began to catch a glimpse of the objects' histories. "Hubble's biggest contribution is enabling us to clock how fast the Magellanic clouds are moving," said Gurtina Besla, a researcher at the University of Arizona who studies dwarf galaxies. In 2007, Besla overturned conventional wisdom when she suggested that the LMC and SMC were making their first orbit of our galaxy.
"They're moving too fast to have been long-term companions of the Milky Way," Besla said.

She used data from the European Space Agency's Gaia spacecraft to clock smaller, satellite galaxies orbiting the LMC, as well. And, understanding how these galaxies move has helped researchers better calculate the mass of the LMC. Current estimates put the LMC at about 100 billion times as massive as Earth's sun, or a quarter the mass of the Milky Way. Besla said this size means the LMC is about 10 times heavier than previously calculated.

(Excerpt end)

Observation:

When new telescopes provide resolution to individual stars the astronomers could ignore these red shifts implying great distance. When an astronomer says "too fast" the basis for that conclusion is not provided. Is it too fast based on which observational evidence? Both Magellanic Clouds are not "typical" spiral galaxies.

Stars in M31 don't rotate like planets around a star so dark matter was needed to fix that wrong expectation.

Using new telescopes to capture their lateral motion, astronomers concluded they are moving too fast.

Dark matter is proposed to explain unexpected galaxy motions in clusters.

Researchers in 2006 concluded our Magellanic Clouds are probably moving past the Milky Way and are not in orbit as previously assumed. In 2014 the first measurement of the LMC rotation period was taken.

This claim of "too fast" cannot be justified when knowing the surprises in the Magellanic Clouds and no history of observations of individual stars in other galaxies.

## 15.5.1 SMC

A story about motion within SMC, involving the same researcher Besla for LMC, titled:

## Small Magellanic Cloud: A Satellite Dwarf Galaxy Neighbor

Excerpt:

For years, most astronomers assumed that the LMC and SMC had circled our galaxy multiple times. However, in 2007, research suggested that the two galaxies were making their first trip around the Milky Way. By using the Hubble Space Telescope to measure the speed of the galaxies relative to one another, University of Virginia astronomer Nitya Kallivayalil and her colleagues calculated the accurate 3D velocities of the LMC and SMC.
"We found that the velocities of the LMC and SMC are unexpectedly large — almost twice those previously thought," Kallivayalil, then a researcher at the Harvard-Smithsonian Center for Astrophysics, said in a statement. There are two possible explanations for these high-speed galaxies: Either the Milky Way is larger than expected, or the Magellanic clouds are not gravitationally connected to our galaxy but are just passing by.
Additional research from Gurtina Besla, an astronomer at the University of Arizona, revealed that the galaxy pair is indeed bound to the Milky Way, which is more massive than originally thought. At the same time, Besla said that understanding the mass of the Milky Way "completely changes" our picture of the history of the Magellanic clouds.

Astronomers have also used the European Southern Observatory (ESO) Very Large Telescope (VLT) in Chile, along with other instruments, to identify, for the first time, an isolated neutron star with low-energy magnetic field located outside the Milky Way. Researchers discovered the star at the center of a ring of gas expanding slowly amid other rings of gas and dust that were left behind after a supernova in the SMC.

"If you look for a point source, it doesn't get much better than when the universe quite literally draws a circle around it to show you where to look," lead author Frédéric Vogt, an ESO fellow in Chile, said in a statement.

Astronomers have also mapped cosmic rays from the SMC and its neighboring LMC in unprecedented detail, thanks to data from the Murchison Widefield Array radio telescope in Australia. Cosmic rays are charged particles emitted from supernova explosions that interact with magnetic fields, creating radiation visible to radio telescopes. Identifying the rays allows scientists to estimate the number of new stars being formed. Researchers estimated that the rate of star formation in the SMC is roughly equivalent to one new star with the mass of our sun every 40 years, according to a statement from the International Centre for Radio Astronomy Research.

"We were able to observe a powerful outflow of hydrogen gas from the Small Magellanic Cloud," lead researcher Naomi McClure-Griffiths, of the Australian National University, said in a statement. Gas flowing out of the galaxy is lost to the star-formation process; once all of the gas is gone, no new stars can form. Galaxies that no longer form stars "gradually fade away into oblivion," she said.

As the SMC loses gas and slows in its star formation, it will eventually fall into more-massive objects, such as our own galaxy.

(Excerpt end)

Observation:

The "outflow of hydrogen gas" is probably a stream of protons, aka, hydrogen ions.
They must detect a plasma filament, and, as is often done, call it gas.

## 15.5 Bridge between the Magellanic Clouds.

The 2017 story was titled:

For The First Time, Physicists Have Observed a Giant Magnetic 'Bridge' Between Galaxies

Excerpt:

"In general, we don't know how such vast magnetic fields are generated, nor how these large-scale magnetic fields affect galaxy formation and evolution," said Kaczmarek.

(Excerpt end)

Observation:

Magnetic fields are created by an electric current.

## 15.6 Cartwheel galaxy

The 2006 NASA story was titled:
NASA - Cartwheel Galaxy Makes Waves in New NASA Image

Excerpt from the story:

new image from NASA's Galaxy Evolution Explorer completes a multi-wavelength, neon-colored portrait of the enormous Cartwheel galaxy after a smaller galaxy plunged through it, triggering ripples of sudden, brief star formation.

Image right: This false-color composite image shows the Cartwheel galaxy. Although astronomers have not identified exactly which galaxy collided with the Cartwheel, two of three candidate galaxies can be seen in this image to the bottom left of the ring, one as a neon blob and the other as a green spiral.

The false-color composite image shows the Cartwheel galaxy as seen by Galaxy Evolution Explorer in ultraviolet light (blue); the Hubble Space Telescope in visible light (green); the Spitzer Space Telescope in infrared (red); and the Chandra X-ray Observatory (purple).

"The dramatic plunge has left the Cartwheel galaxy with a crisp, bright ring around a zone of relative calm," said astronomer Phil Appleton of the California Institute of Technology, Pasadena, Calif. "Usually a galaxy is brighter toward the center, but the ultraviolet view indicates the collision actually smoothed out the interior of the galaxy, concentrating older stars and dust into the inner regions. It's like the calm after the storm of star formation."

The outer ring, which is bigger than the entire Milky Way galaxy, appears blue and violet in the image.

Recently-observed features include concentric rings rippling out from the impact area in a series of star formation waves, ending in the outermost ring. "It's like dropping a stone into a pond, only in this case, the pond is the galaxy, and the wave is the compression of gas," said Appleton. "Each wave represents a burst of star formation, with the youngest stars found in the outer ring."

Previously, scientists believed the ring marked the outermost edge of the galaxy, but the latest Galaxy Evolution Explorer observations detect a faint disk, not visible in this image, that extends to twice the diameter of the ring. This means the Cartwheel is a monstrous 2.5 times the size of the Milky Way.

Most galaxies have only one or two bright X-ray sources, usually associated with gas falling onto a black hole from a companion star. The Cartwheel has a dozen. Appleton said that makes sense, because black holes thrive in areas where massive stars are forming and dying fast.

The Cartwheel galaxy is one of the brightest ultraviolet energy sources in the local universe. In some visible-light images, it appears to have spokes.

(Excerpt end)

Observation:
The reasons why this galaxy must be electrically active are its unique features beyond its ring and dimly luminous plasma filaments:

a) It has a dozen X-ray point sources.
As they are probably scattered outside the central core and one can assume they are individual plasmoids. They are certainly not black holes.

b) There are visible spokes to the outer ring.
Those are probably luminous plasma filaments.

This large ring with such visible details is definitely electrically active because the filaments are luminous. The luminous structure is not densely packed stars.

Clicking on the image in the story zooms in for a better view.

Here are more images including the Chandra image with those individual X-ray sources, several near the core, with others in the ring.

Everything observed is electrical.

The galaxy is bright in ultraviolet.

No spectrum for any region is provided to observe emission lines coming from ions capturing electrons (the lines identify the elements), or to observe synchrotron radiation coming from electric currents bending their path by a magnetic field.

## 15.7 Rings and arcs

These rings and arcs are often around giant elliptical galaxies, but not other galaxy types.

Author has a collection on a web page titled:

**Collection of Rings and Distortions, With Images**

The page offers links to descriptions and images.

The following objects are in the collection. For more information and images from online sources, please use the References page at the end of this book to get that web page providing those links.

These objects are in no particular order.

Hoag's Object is a known ring galaxy.

From Wikipedia:

In the initial announcement of his discovery, Hoag proposed the hypothesis that the visible ring was a product of gravitational lensing. This idea was later discarded because the nucleus and the ring have the same redshift, and because more advanced telescopes revealed the knotty structure of the ring, something that would not be visible if the ring were the product of gravitational lensing.

(Excerpt end)

The criteria about knots is not applied consistently because many other rings show knots but are still claimed to be a distortion by a lens.

There are other ring galaxies

Cartwheel Galaxy
Impressive ring with large radius, with faint spokes

Cartwheel galaxy was the section 15.6 above.

Here is another ring galaxy:

AM 0644-741 or Lindsay-Shapley Ring

It looks like Hoag's Object but the core is left of center

Both X-ray and optical images in References

M94 - has an inner ring and an outer ring of stars

SN1987A - ring with nodes in Chandra image

First image has distant view with a fainter outer ring

Second image is zoomed

Here are rings and arcs claimed to be distortions.

They are in no particular order.

Einstein Cross or QSO 2237+0305 - four nodes around a central node; claimed to be four lensed images of one quasar, the one in the middle.

The simpler explanation is five quasars in close proximity with a possible coarse alignment. Halton Arp observed quasars with the same red shift in pairs.


RXJ1131-1231 - more complex Hoag's object

SDSS J0146-0929 - multiple arcs as parts of a circle

Antennae galaxies - a distant view shows two arcs

Abell 2261 - arc at 9 o'clock; some documents mention knots in Abell 2261.

Abell 383 - long arc at 4 o'clock ending with a knot; beyond that at 5 o'clock is elliptical with distorted circle.

Abell S1063 - several arcs around the elliptical

Abell 1413 - arc at 10 o'clock

Abell 2218 - multiple arcs around the large elliptical at left; filaments around elliptical near top right

Abell 1689 - faint arcs at 2 and 4 o'clock

Abell 2261 - arc at 9 o'clock from large elliptical

Abell 2390 - in X-ray both images show arcs

Abell 2667 large arc to left of large elliptical, the bottom of the arc has a knot

Abell 1835 could have a jet at 1 o'clock to the very bright central object

Abell 370 - large arc at 2 o'clock to central elliptical; the bottom of the arc has a knot

SDSS J1004+4112 - the central bright object is claimed to be a lensed quasar (which don't have jets) but it looks like an elliptical with a long jet at about 10 o'clock; there is an arc with knots at 4 o'clock with another small arc above it; there is also an arc at 2 o'clock that crosses a distant galaxy - quite the coincidence!
When zooming into an image there is another elliptical some distance away from the main one at about 9 o'clock. Oddly this one has an arc with knots at 10 o'clock. Near the right edge is either: two galaxies are merging or one is splitting. Below that is two spirals with long tails. This is a very interesting cluster.

A link in References has image with annotations

SDP.81 - almost a complete circle with a knot inside the circle

LRG 3-757 - the horse shoe Einstein ring

RCS2 032727-132623 - a very large arc around the central elliptical; odd filaments at 4 and 12 o'clock.

The Twin Quasar - adjacent identical quasars are not allowed so they must be an illusion; there are 2 adjacent bright objects in the image; the right one in the pair might have a ring; at about 2 o'clock from the quasar pair at some distance is a reddish ring galaxy.

Abell 1185 - in APOD image: at the left is an elliptical with a very long jet with 2 disorganized nodes; this image is here for that interesting jet, not as a lens.

Abell 2744 Wikipedia has a terrible image with false reds and blues; zoom into the image below; the lower elliptical to the right of a bright spiral has long jets at 2 and 6 o'clock; there is a distant filament at 4 o'clock; the elliptical at 10 o'clock to the one noticeable bright yellow star has filaments at 5 and 10 o'clock

Abell 2261 - arc at 9 o'clock to the BCG

El Gordo - filament at about 1 o'clock

J1531+3414 - cluster with much of interest
double ellipticals at center of very large diameter ring
right one has jet at 2 o'clock but jet is a string of pearls

Second link has inset of jet or a string of pearls

Third link is zoomed to show more details.

ZwCl0024+1652 (CL0024+17 for short) is a galaxy cluster with a ghostly ring of dark matter
Some images show a wide diffuse ring around the ellipticals at center

Second link has an image having the overlay of the dark matter map

Another image shows many arcs around the central ellipticals

PLCK_G308.3-20.2 - called a colossal cluster

SPT 0615- furthest galaxy ever imaged by means of a lens
But the image does not show a large galaxy to serve as a claimed lens, nor does it identify the distorted galaxy.

From the link: '"The galaxy is located towards the upper left, to the right of the group of two stars and one galaxy"

A spiral galaxy is to the far left of the very bright foreground star, above 2 foreground stars. This is not the upper left.

In this image the claimed distant galaxy must be too dim to see. Also in the upper left there are no large galaxies to serve as the lens.
I can only assume this happened here: The observed brightness (though dim) is too bright for its extreme distance calculated by red shift so a lens is the explanation for the anomaly.

SDSS J103842.59+484917.7 - Cheshire Cat

link to a story with images in different wavelengths

Summary of collection of images:

Many of these distant galaxy clusters appear to exhibit much electrical activity as arcs. Perhaps other readers will find even more features than the obvious ones after zooming.

The fine rings and arcs appear part of a circular plasma filament around a large elliptical galaxy where the visible portion is in arc mode and the rest is in dark mode.

There must be more of these "distortions" than I could find, or perhaps Hubble will reveal more images.

# 16 Inter-Galactic Behaviors

Dark matter arose in the 1930's when insufficient data resulted in the conclusion the observational data could not match expectations.

Excerpt from Wikipedia about Fritz Zwicky:

While examining the Coma galaxy cluster in 1933, Zwicky was the first to use the virial theorem to discover the existence of a gravitational anomaly, which he termed dunkle (kalt) Materie 'dark matter'. The gravitational anomaly surfaced due to the excessive rotational velocity of luminous matter compared to the calculated gravitational attraction within the cluster. He calculated the gravitational mass of the galaxies within the cluster from the observed rotational velocities and obtained a value at least 400 times greater than expected from their luminosity. The same calculation today shows a smaller factor, based on greater values for the mass of luminous material; but it is still clear that the great majority of matter was correctly inferred to be dark.

(Excerpt end)

Observation:
With modern numbers, the factor dropped from 400 to "great majority"

In 1933, Zwicky was believed though lacking in accurate data to prove his conclusion.

Some supposed collisions will be in stories in this section.

Collisions of galaxies have no basis. As noted in the first book, no distant galaxy ever had its proper motion measured. To propose a collision there must be evidence for motion implying either a collision or a process of fissioning, the opposite interaction.

## 16.1 Laniakeia

Excerpt from Wikipedia:

The Local Supercluster, Local SCl, or Laniakea Supercluster (Laniakea, Hawaiian for open skies or immense heaven), is the galaxy supercluster that is home to the Milky Way and approximately 100,000 other nearby galaxies. It was defined in September 2014, when a group of astronomers including R. Brent Tully of the University of Hawaii, Hélène Courtois of the University of Lyon, Yehuda Hoffman of the Hebrew University of Jerusalem, and Daniel Pomarède of CEA Université Paris-Saclay published a new way of defining superclusters according to the relative velocities of galaxies. The new definition of the local supercluster subsumes the prior defined local supercluster, the Virgo Supercluster, as an appendage.

Follow-up studies suggest that the Local Supercluster is not gravitationally bound; it will disperse rather than continue to maintain itself as an overdensity relative to surrounding areas.

(Excerpt end)

Observation:

"Local Supercluster is not gravitationally bound" so the only other force at play here is the magnetic force which can affect plasma and objects with electric fields.

## 16.2 Inter-galactic magnetic fields

Donald Scott: No Magnetic Universe w/o Electric Currents | Space News

Excerpt from header for the video:

It's important to note, as we've done in numerous past episodes, that mainstream cosmologists did not predict the so-called magnetic universe. In fact, a telling glimpse into the prevailing thoughts on cosmic magnetism can be found in the 1999 NASA web item entitled, 'Do magnetic fields exist throughout space?' It states, "On the cosmological scale, there is no data to suggest that magnetic fields are present. They certainly are not important in the dynamics of the universe for any reasonable range of field strengths consistent with present observational constraints." However, today, not only do mainstream cosmologists view cosmic magnetism as important, some astrophysicists are counting on mysterious magnetism to resolve ad hoc some of the biggest evidentiary problems for the Big Bang hypothesis

Excerpt end)

Observation:

The video explains the electrical origin of magnetic fields on the inter-galactic scale.

## 16.2 Pandora Cluster

The Chandra story title:

Abell 2744: Pandora's Cluster Revealed

According to Chandra, Abell 2744 is:

"A complex collision of at least four galaxy clusters is captured in this new image."

Data from NASA's Chandra X-ray Observatory are colored red, showing gas with temperatures of millions of degrees.

(Excerpt end)

Observation:

The Chandra description has much nonsense about fictitious dark matter.

The Chandra page has the image available in different wave lengths.

Select Composite for the image having blue to indicate where fictitious dark matter is supposedly present

Select X-Ray for the image showing these ridiculous temperatures.

There are several red, diffused "clouds" of strong X-ray emissions.

In non-electrical cosmology, X-rays can come from only a ridiculous impossible temperature, like in this story.

In electrical cosmology, X-rays are at the high frequency end of synchrotron radiation.

When either electrons or an electrical current bend their path by a magnetic field, like in a synchrotron, synchrotron radiation results.

The likely cause of X-rays in a diffused cloud is many individual electrons spiraling due to a magnetic field, perhaps around a filament.

It is notable the X-ray image shows no X-ray point sources. Those would be plasmoids like at a galaxy center, but there are none.

Instead the composite image overlay identifies several objects in optical as part of the claimed clusters.

Having no basis for a galaxy collision, they fall back to a cluster collision. The image in optical shows no apparent clusters of objects but instead an apparent random distribution.

Chandra has other images showing galaxy clusters with a similar X-ray background.

## 16.3 Phoenix Cluster

The Chandra headline: A Weakened Black Hole Allows Its Galaxy to Awaken

Excerpt:

Astronomers have confirmed the first example of a galaxy cluster where large numbers of stars are being born at its core. Using data from NASA space telescopes and a National Science Foundation radio observatory, researchers have gathered new details about how the most massive black holes in the universe affect their host galaxies. Galaxy clusters are the largest structures in the cosmos that are held together by gravity, consisting of hundreds or thousands of galaxies embedded in hot gas, as well as invisible dark matter. The largest supermassive black holes known are in galaxies at the centers of these clusters.
For decades, astronomers have looked for galaxy clusters containing rich nurseries of stars in their central galaxies. Instead, they found powerful, giant black holes pumping out energy through jets of high-energy particles and keeping the gas too warm to form many stars.
Now, scientists have compelling evidence for a galaxy cluster where stars are forming at a furious rate, apparently linked to a less effective black hole in its center. In this unique cluster, the jets from the central black hole instead appear to be aiding in the formation of stars. Researchers used new data from NASA's Chandra X-ray Observatory and Hubble Space Telescope, and the NSF's Karl Jansky Very Large Array (VLA) to build on previous observations of this cluster.

"This is a phenomenon that astronomers had been trying to find for a long time," said Michael McDonald, astronomer at the Massachusetts Institute of Technology (MIT), who led the study. "This cluster demonstrates that, in some instances, the energetic output from a black hole can actually enhance cooling, leading to dramatic consequences."

The black hole is in the center of a galaxy cluster called the Phoenix Cluster, located about 5.8 billion light years from Earth in the Phoenix constellation. The large galaxy hosting the black hole is surrounded by hot gas with temperatures of millions of degrees. The mass of this gas, equivalent to trillions of Suns, is several times greater than the combined mass of all the galaxies in the cluster.

This hot gas loses energy as it glows in X-rays, which should cause it to cool until it can form large numbers of stars. However, in all other observed galaxy clusters, bursts of energy driven by such a black hole keep most of the hot gas from cooling, preventing widespread star birth.

"Imagine running an air-conditioner in your house on a hot day, but then starting a wood fire. Your living room can't properly cool down until you put out the fire," said co-author Brian McNamara of the University of Waterloo in Canada. "Similarly, when a black hole's heating ability is turned off in a galaxy cluster, the gas can then cool."

Evidence for rapid star formation in the Phoenix Cluster was previously reported in 2012 by a team led by McDonald. But deeper observations were required to learn details about the central black hole's role in the rebirth of stars in the central galaxy, and how that might change in the future.

By combining long observations in X-ray, optical, and radio light, the researchers gained a ten-fold improvement in the data quality compared to previous observations. The new Chandra data reveal that hot gas is cooling nearly at the rate expected in the absence of energy injected by a black hole. The new Hubble data show that about 10 billion solar masses of cool gas are located along filaments leading towards the black hole, and young stars are forming from this cool gas at a rate of about 500 solar masses per year. By comparison, stars are forming in the Milky Way galaxy at a rate of about one solar mass per year.

The VLA radio data reveal jets blasting out from the vicinity of the central black hole. These jets likely inflated bubbles in the hot gas that are detected in the Chandra data. Both the jets and bubbles are evidence of past rapid growth of the black hole. Early in this growth, the black hole may have been undersized, compared to the mass of its host galaxy, which would allow rapid cooling to go unchecked.

"In the past, outbursts from the undersized black hole may have simply been too weak to heat its surroundings, allowing hot gas to start cooling," said co-author Matthew Bayliss, who was a researcher at MIT during this study, but has recently joined the faculty at the University of Cincinnati. "But as the black hole has grown more massive and more powerful, its influence has been increasing."

The cooling can continue when the gas is carried away from the center of the cluster by the black hole's outbursts. At a greater distance from the heating influence of the black hole, the gas cools faster than it can fall back towards the center of the cluster. This scenario explains the observation that cool gas is located around the borders of the cavities, based on a comparison of the Chandra and Hubble data. Eventually the outburst will generate enough turbulence, sound waves and shock waves (similar to the sonic booms produced by supersonic aircraft) to provide sources of heat and prevent further cooling. This will continue until the outburst ceases and the build-up of cool gas can recommence. The whole cycle may then repeat.
"These results show that the black hole has temporarily been assisting in the formation of stars, but when it strengthens its effects will start to mimic those of black holes in other clusters, stifling more star birth," said co-author Mark Voit of Michigan State University in East Lansing, Michigan.
The lack of similar objects shows that clusters and their enormous black holes pass through the rapid star formation phase relatively quickly.

(Excerpt end)

Observation:
There is so much wrong in this explanation.

a) "galaxies embedded in hot gas, as well as invisible dark matter"

"Hot gas" always means plasma.

"invisible dark matter" simply does not exist.

b) "10 billion solar masses of cool gas are located along filaments" means filaments of mostly protons, which are not capturing electrons to generate UV emission lines to suggest something to perceive as "hot." The basis for this mass calculation is unknown.

c) "giant black holes pumping out jets" is nonsense.

Black holes prevent anything from escaping. Plasmoids are known to have jets with M87 known for the material's high velocity.

d) "These jets likely inflated bubbles in the hot gas that are detected in the Chandra data" means there are clouds of plasma generating synchrotron radiation being observed by Chandra.

e) "undersized black hole" is also nonsense. There is no method to measure the mass or size of a black hole. A black hole is supposed to be a singularity, or a mass in zero volume at infinite density, or impossible. A black hole is just a theoretical object required for an X-ray source when a plasmoid is ignored as the most likely explanation.

Any assigned characteristics for a black hole have no basis. An arbitrary amount of mass within the fictitious black hole is often assigned with no evidence to support the value selected.

Any jet behavior can be explained by a plasmoid because they are known to exhibit that behavior, like the plasmoid in M87.

## 16.5 Spiral Galaxies in Collision

This is the title for the Astronomy Picture of the Day on 2004 November 21.

Excerpt from its description:

Billions of years from now, only one of these two galaxies will remain. Until then, spiral galaxies NGC 2207 and IC 2163 will slowly pull each other apart, creating tides of matter, sheets of shocked gas, lanes of dark dust, bursts of star formation, and streams of cast-away stars. Astronomers predict that NGC 2207, the larger galaxy on the left, will eventually incorporate IC 2163, the smaller galaxy on the right. In the most recent encounter that peaked 40 million years ago, the smaller galaxy is swinging around counter-clockwise, and is now slightly behind the larger galaxy. The space between stars is so vast that when galaxies collide, the stars in them usually do not collide.

(Excerpt end)

First of all, neither of the two galaxies has a measured 3-dimensional velocity.

Without that detail, it is impossible to determine convergence or divergence. Assigning a time for an earlier "encounter peak" or a time for any future result cannot be justified.

"sheets of shocked gas" must be a reference to plasma.

APOD provides only "a brief explanation written by a professional astronomer."

For comparison, Chandra also has a story on this "collision" of 2 spiral galaxies.

Excerpt from Chandra:

NGC 2207 and IC 2163 are two spiral galaxies in the process of merging.

This pair contains a large collection of super bright X-ray objects called "ultraluminous X-ray sources" (ULXs).

Astronomers have found evidence for three supernova explosions within this pair in the past 15 years.

A new composite image of the system contains X-rays from Chandra (pink) along with optical and infrared data.

ULXs have far brighter X-rays than most "normal" X-ray binaries. The true nature of ULXs is still debated, but they are likely a peculiar type of X-ray binary. The black holes in some ULXs may be heavier than stellar mass black holes and could represent a hypothesized, but as yet unconfirmed, intermediate-mass category of black holes.

This composite image of NGC 2207 and IC 2163 contains Chandra data in pink, optical light data from the Hubble Space Telescope in red, green, and blue (appearing as blue, white, orange, and brown), and infrared data from the Spitzer Space Telescope in red.

The new Chandra image contains about five times more observing time than previous efforts to study ULXs in this galaxy pair. Scientists now tally a total of 28 ULXs between NGC 2207 and IC 2163. Twelve of these vary over a span of several years, including seven that were not detected before because they were in a "quiet" phase during earlier observations.

The scientists involved in studying this system note that there is a strong correlation between the number of X-ray sources in different regions of the galaxies and the rate at which stars are forming in these regions. The composite image shows this correlation through X-ray sources concentrated in the spiral arms of the galaxies, where large amounts of stars are known to be forming. This correlation also suggests that the companion star in the binary systems is young and massive.

(Excerpt end)

Observation:

Chandra's page includes tabs to select optical, like APOD, and also X-ray and Infrared.

The X-ray image reveals those "ULXs [whose true nature] is still debated."

Those 28 ULXs in the image are seen in optical, and X-ray.

Infrared can be obscured, unlike X-ray, so its image is not always consistent.

That broad frequency spectrum from each object matches synchrotron radiation.

The 28 objects are all plasmoids.

In the big bang cosmology, every X-ray point source is a black hole. M87 demonstrated they are often plasmoids.

On May 21, I posted "Plasmoid Ejection" which observed the Birkelund current pair with a pinch or bend appears the combination required for ejecting plasmoids. That combination is not restricted to only a galaxy core.

Both galaxies in this claimed "collision" are spiral galaxies. I did not verify the 28 but whatever the number they seem mostly in the spiral arms.

The pinch at the core of IC 2163 at the right is more intense in X-ray than the core of NGC 2207 at the left.

Perhaps, these are coincidences but around the IC 2163 core:

At 4 o'clock and 10 o'clock, there is a dimmer pair of X-ray points nearby, and a brighter pair of point pairs further away.

Arp famously observed ejected pairs of quasars which are just plasmoids and surrounded by clouds of metal atoms. Seyfert galaxies (Arp's parent of quasars) provide the metals. Neither of these galaxies are Seyferts so no quasars.

Chandra images are always entertaining. X-rays are possible only with synchrotron radiation. Thermal radiation never exceeds ultraviolet.

The infrared in the image from Spitzer is at the lower end of synchrotron radiation.

Apparently Chandra has been observing this collision awhile to note 12 ULXs have varied "over a span of several years."

In a laboratory, plasmoids don't last long. On the galactic scale they last much longer.

The collision story with many ULXs is much more exciting than 2 electrically active spiral galaxies.

## 16.6 NGC 1316: After Galaxies Collide

This is the title for the Astronomy Picture of the Day on 2005 April 4 and again on 2017 September 9.

Excerpt from its 2005 description:

How did this strange-looking galaxy form? Astronomers turn detectives when trying to figure out the cause of unusual jumbles of stars, gas, and dust like NGC 1316. A preliminary inspection indicates that NGC 1316 is an enormous elliptical galaxy that includes dark dust lanes usually found in a spiral. The above image taken by the Hubble Space Telescope shows details, however, that help in reconstructing the history of this gigantic jumble. Close inspection finds fewer low mass globular clusters of stars toward NGC 1316's center. Such an effect is expected in galaxies that have undergone collisions or merging with other galaxies in the past few billion years. After such collisions, many star clusters would be destroyed in the dense galactic center. The dark knots and lanes of dust indicate that one or more of the devoured galaxies were spiral galaxies. NGC 1316 spans about 60,000 light years and lies about 75 million light years away toward the constellation of the Furnace.

(Excerpt end)

Observation:

Astronomers see "unusual jumbles of stars, gas, and dust " and HST also provided detailed images of globular clusters.

This galaxy suggests astronomers are just clueless about elliptical galaxies.

There is no attempt to understand the galaxy. Instead the focus is on its periphery.

They claim this giant elliptical has absorbed other galaxies in the past.

Apparently, those visible dusy lanes on the OUTSIDE are considered remnants of those merged galaxies, now gone, whose GC stars went to the INSIDE, where they were "destroyed in the dense galactic center."


APOD provides only "a brief explanation written by a professional astronomer."

This explanation is suspicious after viewing the image.


APOD page had this same title on Sept 9, 2017.

The later APOD offers a wider view revealing there is a large, spiral galaxy, NGC 131, very near the elliptical NGC 1316 shown in APOD.

Excerpt from second APOD:

An example of violence on a cosmic scale, enormous elliptical galaxy NGC 1316 lies about 75 million light-years away toward Fornax. Investigating the startling sight, astronomers suspect the giant galaxy of colliding with smaller neighbor NGC 1317 seen just above, causing far flung loops and shells of stars. Light from their close encounter would have reached Earth some 100 million years ago. In the deep, sharp image, the central regions of NGC 1316 and NGC 1317 appear separated by over 100,000 light-years. Complex dust lanes visible within also indicate that NGC 1316 is itself the result of a merger of galaxies in the distant past. Found on the outskirts of the Fornax galaxy cluster, NGC 1316 is known as Fornax A. One of the visually brightest of the Fornax cluster galaxies it is one of the strongest and largest radio sources with radio emission extending well beyond this telescopic field-of-view, over several degrees on the sky.

(Excerpt end)

Observation:

APOD neglected to mention NGC 1316 "is one of the strongest and largest radio sources [covering 6 degrees on the sky]."

There is an unjustified conclusion here: "example of violence on a cosmic scale." There is nothing here to justify that "example" except for the other galaxy.

Unless both galaxies have a Cepheid, both their distances are unknown to know whether they are close only by line of sight or are physically far apart.

APOD provides a link to a study of the globular clusters around NGC 1316:

Excerpt:

Recent observations of globular clusters (GCs) in intermediate-age (2--4 Gyr old), early-type merger remnants have provided the hitherto "missing link" between young merger remnants and 'normal' elliptical galaxies in the form of a GC subsystem with colors and luminosities consistent with population synthesis model predictions for those ages and ~ solar metallicity. Here we present new, deep observations of the GC system of the intermediate-age merger remnant NGC 1316, using the ACS camera aboard Hubble Space Telescope, which allowed us to create luminosity functions (LFs) as a function of galactocentric radius. We find that the inner 50% of the 'red' GC system shows a clear turnover in its LF, at about 1 mag fainter than that of the `old' blue GCs. This constitutes direct, dynamical evidence that metal-rich GC populations formed during a gas-rich merger can evolve into the 'red', metal-rich GC populations that are ubiquitous in 'normal' giant ellipticals.

(Excerpt end)

Observation:

The NGC 1516 globular cluster star populations do not conform to "normal" elliptical galaxies in the form of a GC subsystem"

These deviations conclude: "This constitutes direct, dynamical evidence" of their claimed merger of different globular clusters from another galaxy.

Cosmologists have no historical evidence for:

a) how specific metallicity changes by age.

We have no history of metallicity changes in our Sun, let alone any other star. "solar metallicity" is mentioned.

Perhaps GC stars evolve differently than stars in spiral galaxies.

b) whether metallicity in a GC is related only to its parent galaxy.

The assumption here is: if the GC metallicity does not match that expected, then the GC came from another galaxy.

There is no evidence for that direct connection. It just assumes stars can never have an unexpected change in metallicity, because it is "well established" the stellar surface is driven by metals in the dust cloud at the time of the star's formation.

Oddly, supernovae are assumed to distribute metals into space but that interstellar "pollution" is ignored in these metallicity conclusions.

Sometimes, a metallicity analysis requires the frequency of supernovae in the vicinity to anticipate that pollution of metals.

One can be astounded by how cosmologists make baseless assumptions lacking in evidence, and then make unfounded conclusions, like a collision, on the vague data.

Stories about galaxies like this one, which is a "startling sight," only reveal how cosmology lacks a foundation in science by evidence. Guesses about evolution of stars in globular clusters lead to a laughable conclusion of "violence on a cosmic scale."

## 16.7 Galactic Collision in Abell 1185

This is the title for the Astronomy Picture of the Day on 2005 November 22.

Excerpt from its description:

What is a guitar doing in a cluster of galaxies? Colliding. Clusters of galaxies are sometimes packed so tight that the galaxies that compose them collide. A prominent example occurs on the left of the above image of the rich cluster of galaxies Abell 1185. There at least two galaxies, cataloged as Arp 105 and dubbed The Guitar for their familiar appearance, are pulling each other apart gravitationally. Most of Abell 1185's hundreds of galaxies are elliptical galaxies, although spiral, lenticular, and irregular galaxies are all clearly evident. Many of the spots on the above image are fully galaxies themselves containing billions of stars, but some spots are foreground stars in our own Milky Way Galaxy. Recent observations of Abell 1185 have found unusual globular clusters of stars that appear to belong only to the galaxy cluster and not to any individual galaxy. Abell 1185 spans about one million light years and lies 400 million light years distant.
(Excerpt end)

Observation:

The "guitar" is a striking vertical alignment at the left.

A link in References hast the image from APOD but with the "guitar" at the bottom.

Wikipedia identifies only two NGC numbers for this Abell 1185 cluster. NGC 3558 and 3561.
Both offer very little information.

NGC 3558 data: Its galaxy type is uncertain: E2, S0p

It is either an elliptical or a lenticular which is peculiar.

There are 2 elliptical galaxies on the opposite side from the guitar. NGC 3558 must be one while the other in not named.

This NGC 3561 is actually a pair of A and B galaxies:
galaxy: type
A: Sa - spiral
B: E5 - elliptical

Apparently, this pair is considered as too close than "allowed" so they "must be" colliding.

Without reliable data for distance and velocity for everything, that conclusion cannot be justified, when based on only the line of sight.


The Wikipedia NGC 3561 image is the best for detail.

The 2 galaxies seem further apart in this image.

Excerpt from Wikipedia:

NGC 3561, also known as Arp 105, is a pair of interacting galaxies NGC 3561A and NGC 3561B within the galaxy cluster Abell 1185 in Ursa Major. Its common name is "the Guitar" and contains a small tidal dwarf galaxy known as Ambartsumian's Knot that is believed to be the remnant of the extensive tidal tail pulled out of one of the galaxies.

(Excerpt end)

Observation:

Calling the "knot" a small dwarf galaxy does not seem justified.

There is another way to interpret this "collision":

NGC 3561B is a giant elliptical galaxy like M87, which is type E0.
M87 is known for its energetic jets in opposing directions from the plasmoid at the core. One jet is more than the other.

NGC 3561B has opposing non-symmetrical jets.

Ambartsumian's Knot could be a very short jet in one direction.
The "guitar neck" could be a long jet in the opposite direction.

Collisions should never be proposed unless there is proper evidence. Proximity alone is not sufficient.

A more likely explanation than a collision is:

a) there are opposing jets from 3561B,
b) both the distance and amount of material are different for the jets.

c) A small spiral galaxy 3561A is in the line of sight, in front of the jet, on that side of 3561B.

d) the knot is not the remnants of a dwarf galaxy,

e) though the distant jet has substantial material, it seems to have no NGC number and no name (unlike the knot).

The image quality does not justify observing a second jet behind 3561A, almost through a distant star on that side.

With no other data, especially the relative distances of both galaxies, that explanation matches the observation.

## 16.8 Filaments in Cosmic Web

The story is titled:

Faint Filaments of Universe-Spanning 'Cosmic Web' Finally Found

Excerpt:

The scientists focused on the SSA22 Protocluster, which lies about 12 billion light-years away from Earth in the constellation Aquarius. A protocluster is a group of hundreds to thousands of galaxies that are beginning to form a galaxy cluster, the largest structures held together by gravity in the universe.
Using the Multi Unit Spectroscopic Explorer instrument on the European Southern Observatory's Very Large Telescope in Chile, the researchers detected and mapped light emitted by hydrogen gas excited by ultraviolet rays from galaxies within the protocluster. The instrument was designed to scan wide swaths of the sky to spot the faintest structures known.
Essentially, the core of the protocluster acted like a flashlight to help illuminate the otherwise elusive filaments of the cosmic web.
Gas around the young galaxies in the protocluster was arranged in long filaments extending over more than 3.25 million light-years, the scientists found in the new study, which was published online today (Oct. 3) in the journal Science.

These strands are the brightest threads of the cosmic web found yet, but they're still quite dim. The emission levels from the very outskirts of the filaments are as low as 5% that of the ambient background light from the rest of the sky, said astrophysicist Erika Hamden at the University of Arizona in Tucson, who wrote a commentary piece about the new study in the same issue of Science.

"These gaseous structures were predicted for years theoretically, but astronomers have struggled to map them directly," Umehata said. "Our work shows that mapping cosmic web filaments is now possible, which means that we have obtained a novel tool to understand the formation of galaxies and supermassive black holes."

These filaments are likely feeding gas to the protocluster. "They are thought to fuel the intense activity seen in galaxies — star formation and the growth of supermassive black holes," Umehata said. "Thus, our research adds a key piece to understand how galaxies and supermassive black holes acquire their fuels."

Observation:

Of course this "flashlight" is complete nonsense.

These are plasma filaments with moving protons and electrons.

In the event a proton captures an electron, a hydrogen atom is created, resulting in the emission of the Lyman-alpha emission line, whose wave length is in ultraviolet.

These filaments are essentially an electric current spanning the great distance between galaxies and even between clusters.

## 16.9 Intracluster Medium

The RAS posted a story titled:

Stormy cluster weather could unleash black hole power and explain lack of cosmic cooling

Excerpt:

Typical clusters of galaxies have several thousand member galaxies, which can be very different to our own Milky Way and vary in size and shape. These systems are embedded in very hot gas known as the intracluster medium (ICM), all of which live in an unseen halo of so-called 'dark matter." A large number of galaxies have supermassive black holes in their centres, and these often have high speed jets of material stretching over thousands of light years that can inflate very hot lobes in the ICM.

(Excerpt end)

Observation:

The ICM is plasma, not "hot gas."

When charged particles in motion change their path the result is synchrotron radiation. Depending on the particle velocity, the peak frequency can range from radio to X-ray.

The story does not mention the frequencies observed to conclude "hot" but the radiation in the ICM is never thermal and always synchrotron.

## 16.9 Energy Bursts

The headline:

Mysterious 'fast radio burst' detected closer to Earth than ever before

The theme of the story:

Most FRBs originate hundreds of millions of light-years away. This one came from inside the Milky Way.

Mysterious "fast radio burst" detected closer to Earth than ever before

Observation:

A Fast Radio Burst must be synchrotron radiation. Radio is at the low frequency end with gamma ray at the high end. This story covers both ends.

There are no atomic emission lines in that frequency range of radio. Thermal radiation ranges from UV down to infrared but never radio.

Wherever the FRB originates, there must be somewhat "slow" plasma particles diverting their path of motion due to a magnetic field.

ncreasing the plasma velocity increases the peak frequency
n synchrotron radiation.

For a radio burst, the source of these slow charged particles
s intermittent.

Excerpt from the study:

This is the first burst with a radio counterpart observed from
a soft γ-ray repeater and it strongly supports models based
on magnetars that have been proposed for extragalactic fast
radio bursts.

(Excerpt end)

Observation:
There is a "gamma ray repeater" here, associated with the
FRB.

When the energy of gamma rays is absorbed by an atom, a
wonder of particle physics occurs - particle-pair production.
This pair is often an electron and positron. The energy
absorbed measured in eV exactly matches the summed
mass of the two particles measured in eV.

As a result, there are two slow moving charged particles.
When their motion is altered by a magnetic field, they could
be the source of synchrotron radiation down in the radio
frequency range.

The radio burst would end as soon as the electron and
positron mutually annihilate.

The story mentions a magnetar.

Excerpt from Wikipedia:

A magnetar is a type of neutron star believed to have an extremely powerful magnetic field (~$10^9$ to $10^{11}$ T, ~$10^{13}$ to $10^{15}$ G). The magnetic field decay powers the emission of high-energy electromagnetic radiation, particularly X-rays and gamma rays. The theory regarding these objects was proposed by Robert Duncan and Christopher Thompson in 1992, but the first recorded burst of gamma rays thought to have been from a magnetar had been detected on March 5, 1979. During the following decade, the magnetar hypothesis became widely accepted as a likely explanation for soft gamma repeaters (SGRs) and anomalous X-ray pulsars (AXPs). On 1 June 2020, astronomers reported narrowing down the source of Fast Radio Bursts (FRBs), which may now plausibly include "compact-object mergers and magnetars arising from normal core collapse supernovae".

(Excerpt end)

Observation:

The magnetar must be an energetic plasmoid with a very high velocity current in its torus to achieve gamma ray emission. This plasmoid must be somewhat unstable to cause of bursts in gamma rays, even to be called repeaters.

There is no defined requirement for which atoms can execute particle-pair production. Apparently whatever atoms are around the plasmoid execute this production.

This story of an impossible solution with a magnetar is similar to black holes.

X-ray point sources had no explanation when ignoring plasma, despite the discovery of synchrotrons in the 1950's.

When X-ray point sources were found in every galaxy, the ridiculous black hole and its impossibly hot accretion disk was the only solution when ignoring a synchrotron.

Recently, gamma ray sources are found in the universe.

Again, when ignoring a synchrotron, a "new" solution was needed so a fictitious magnetar is proposed. This "extremely powerful magnetic field" is probably impossible; this is like the accretion disk whose temperature is actually impossible.
This statement cannot be justified: "The magnetic field decay powers the emission of high-energy electromagnetic radiation, particularly X-rays and gamma rays."

Apparently cosmologists remain unaware that in recent years, gamma rays are detected with extreme lightning bolts here on Earth.

It is very unlikely the observed gamma rays from terrestrial lightning bolts occur by "magnetic field decay."

That lightning observation alone should force cosmologists to reconsider any distant celestial source of gamma rays. Instead we get a magnetar arising from a "core collapse supernovae" as if a tragic collapse is a viable solution for a repeater.

As long as cosmology ignores synchrotron radiation for its known frequency range, mysteries like FRBs and SGRs will not be solved.

# 17 Solar System

There are notable observations about electrical behaviors in our solar system which include important suggestions in the progression of cosmology beyond those steps described in the author's first two books.

Popular cosmology ignores electric and magnetic fields and forces, because of the wrong assumption gravity is crucial and can explain everything. Relativity distorted cosmology.

The 2 significant observations are 1) magnetic fields which require an electric current, and 2) radio emissions which require a slow electric current affected by a magnetic field.

## 17.1 Sun

Noted in section 5 of this book, the Ulysses space probe flew over both poles of the Sun and measured a strong electrical current along the Sun's axis. This was the only possible method for such a measurement.

This is the Sun's electrical connection to the galaxy.

Excerpt from Wikipedia about Ulysses:

discoveries:

Data provided by Ulysses led to the discovery that the Sun's magnetic field interacts with the Solar System in a more complex fashion than previously assumed.
Data provided by Ulysses led to the discovery that dust coming into the Solar System from deep space was 30 times more abundant than previously expected.
In 2007–2008 data provided by Ulysses led to the determination that the magnetic field emanating from the Sun's poles is much weaker than previously observed.
That the solar wind has "grown progressively weaker during the mission and is currently at its weakest since the start of the Space Age."

(Excerpt end)

Excerpt from Wikipedia about solar wind:

The solar wind is a stream of charged particles released from the upper atmosphere of the Sun, called the corona. This plasma mostly consists of electrons, protons and alpha particles with kinetic energy between 0.5 and 10 keV.

The composition of the solar wind plasma also includes a mixture of materials found in the solar plasma: trace amounts of heavy ions and atomic nuclei C, N, O, Ne, Mg, Si, S, and Fe. There are also rarer traces of some other nuclei and isotopes such as P, Ti, Cr, Ni, Fe 54 and 56, and Ni 58,60,62. Embedded within the solar-wind plasma is the interplanetary magnetic field. The solar wind varies in density, temperature and speed over time and over solar latitude and longitude. Its particles can escape the Sun's gravity because of their high energy resulting from the high temperature of the corona, which in turn is a result of the coronal magnetic field.

At a distance of more than a few solar radii from the Sun, the solar wind reaches speeds of 250–750 km/s and is supersonic, meaning it moves faster than the speed of the fast magnetosonic wave. The flow of the solar wind is no longer supersonic at the termination shock. The Voyager 2 spacecraft crossed the shock more than five times between 30 August and 10 December 2007. Voyager 2 crossed the shock about a Tm closer to the Sun than the 13.5 Tm distance where Voyager 1 came upon the termination shock. The spacecraft moved outward through the termination shock into the heliosheath and onward toward the interstellar medium. Other related phenomena include the aurora (northern and southern lights), the plasma tails of comets that always point away from the Sun, and geomagnetic storms that can change the direction of magnetic field lines.

(Excerpt end)

Observation:

The solar wind is essentially the distribution of charged particles past all the bodies in the solar system.

From Wikipedia, it looks like this figure, with this caption.

The heliospheric current sheet results from the influence of the Sun's rotating magnetic field on the plasma in the solar wind
The fusion model of the Sun has many problems, including the corona and solar wind.

Dr. Robitaille, the originator of the liquid metallic hydrogen model for the Sun (extensively described in the second book) has a YouTube video which includes the mechanism for the fast and slow solar wind, titled:

Intercalate Zones in the Sun and Stars!

Excerpt from the header:

Interchelate Zones, Fast Solar Winds, Activity, and Current Flow: Does our Sun have a Charge?

## 17.2 Planet Mercury

Magnetic field

Excerpt from Wikipedia:

Mercury's magnetic field is approximately a magnetic dipole (meaning the field has only two magnetic poles) apparently global, on planet Mercury. Data from Mariner 10 led to its discovery in 1974; the spacecraft measured the field's strength as 1.1% that of Earth's magnetic field. The origin of the magnetic field can be explained by dynamo theory. The magnetic field is strong enough near the bow shock to slow the solar wind, which induces a magnetosphere.

(Excerpt end)

Observation:

The field is strong enough to affect the solar wind.

## 17.3 Planet Venus

The author recalls the confrontation between Velikovsky and Sagan many years ago, when Carl Sagan was at a loss for a valid response to Velikovsky's confirmed prediction of a "hot" Venus.

Apparently, at that instant with no scientific basis, Sagan countered with a "runaway greenhouse effect" just so Sagan could still claim Velikovsky was wrong.

To the detriment of science, this "sound byte" from a celebrity became the accepted explanation for the "hot" Venus.

At the time, and many times since, this nonsense has been debunked because Venus has a dense, high pressure atmosphere which results in a higher temperature, almost regardless of the elements in it. Also, Venus atmosphere is thick with a high albedo, so the surface is not heated by sunlight, which is how a green house works, by surface heating which is not being removed by normal convection.

Excerpt from Wikipedia:

A runaway greenhouse effect involving carbon dioxide and water vapor may have occurred on Venus. In this scenario, early Venus may have had a global ocean if the outgoing thermal radiation was below the Simpson-Nakajima limit but above the moist greenhouse limit. As the brightness of the early Sun increased, the amount of water vapor in the atmosphere increased, increasing the temperature and consequently increasing the evaporation of the ocean, leading eventually to the situation in which the oceans boiled, and all of the water vapor entered the atmosphere. This scenario helps to explain why there is little water vapor in the atmosphere of Venus today. If Venus initially formed with water, the runaway greenhouse effect would have hydrated Venus' stratosphere, and the water would have escaped to space. Some evidence for this scenario comes from the extremely high deuterium to hydrogen ratio in Venus' atmosphere, roughly 150 times that of Earth, since light hydrogen would escape from the atmosphere more readily than its heavier isotope, deuterium. Venus is sufficiently strongly heated by the Sun that water vapor can rise much higher in the atmosphere and be split
into hydrogen and oxygen by ultraviolet light. The hydrogen can then escape from the atmosphere while the oxygen recombines or bonds to iron on the planet's surface.

The deficit of water on Venus due to the runaway greenhouse effect is thought to explain why Venus does not exhibit surface features consistent with plate tectonics, meaning it would be a stagnant lid planet. Carbon dioxide, the dominant greenhouse gas in the current Venusian atmosphere, owes its larger concentration to the weakness of carbon recycling as compared to Earth, where the carbon dioxide emitted from volcanoes is efficiently subducted into the Earth by plate tectonics on geologic time scales through the carbonate-silicate cycle, which requires precipitation to function.
(Excerpt end)

Here is one rebuttal's excerpt:

 The high temperatures there can be almost completely explained by atmospheric pressure – not composition. If 90% of the CO2 in Venus atmosphere was replaced by Nitrogen, it would change temperatures there by only a few tens of degrees.

(Excerpt end)

Observation:

Cosmology must admit Carl Sagan was wrong, and accept Venus does not violate physics with an unjustified "runaway" mechanism.

## 17.4 Planet Earth

There is an electrical connection from Sun to Earth, observed in the aurora.

Excerpt from Wikipedia:

A full understanding of the physical processes which lead to different types of auroras is still incomplete, but the basic cause involves the interaction of the solar wind with the Earth's magnetosphere. The varying intensity of the solar wind produces effects of different magnitudes but includes one or more of the following physical scenarios. A quiescent solar wind flowing past the Earth's magnetosphere steadily interacts with it and can both inject solar wind particles directly onto the geomagnetic field lines that are 'open', as opposed to being 'closed' in the opposite hemisphere, and provide diffusion through the bow shock. It can also cause particles already trapped in the radiation belts to precipitate into the atmosphere. Once particles are lost to the atmosphere from the radiation belts, under quiet conditions, new ones replace them only slowly, and the loss-cone becomes depleted. In the magnetotail, however, particle trajectories seem constantly to reshuffle, probably when the particles cross the very weak magnetic field near the equator. As a result, the flow of electrons in that region is nearly the same in all directions ("isotropic") and assures a steady supply of leaking electrons. The leakage of electrons does not leave the tail positively charged, because each leaked electron lost to the atmosphere is replaced by a low energy electron drawn upward from the ionosphere. Such replacement of "hot" electrons by "cold" ones is in complete accord with the 2nd law of thermodynamics.

The complete process, which also generates an electric ring current around the Earth, is uncertain.
Geomagnetic disturbance from an enhanced solar wind causes distortions of the magnetotail ("magnetic substorms"). These 'substorms' tend to occur after prolonged spells (hours) during which the interplanetary magnetic field has had an appreciable southward component. This leads to a higher rate of interconnection between its field lines and those of Earth. As a result, the solar wind moves magnetic flux (tubes of magnetic field lines, 'locked' together with their resident plasma) from the day side of Earth to the magnetotail, widening the obstacle it presents to the solar wind flow and constricting the tail on the night-side. Ultimately some tail plasma can separate ("magnetic reconnection"); some blobs ("plasmoids") are squeezed downstream and are carried away with the solar wind; others are squeezed toward Earth where their motion feeds strong outbursts of auroras, mainly around midnight ("unloading process"). A geomagnetic storm resulting from greater interaction adds many more particles to the plasma trapped around Earth, also producing enhancement of the "ring current". Occasionally the resulting modification of the Earth's magnetic field can be so strong that it produces auroras visible at middle latitudes, on field lines much closer to the equator than those of the auroral zone.

Acceleration of auroral charged particles invariably accompanies a magnetospheric disturbance that causes an aurora. This mechanism, which is believed to predominantly arise from strong electric fields along the magnetic field or wave-particle interactions, raises the velocity of a particle in the direction of the guiding magnetic field. The pitch angle is thereby decreased and increases the chance of it being precipitated into the atmosphere. Both electromagnetic and electrostatic waves, produced at the time of greater geomagnetic disturbances, make a significant contribution to the energizing processes that sustain an aurora. Particle acceleration provides a complex intermediate process for transferring energy from the solar wind indirectly into the atmosphere.

The details of these phenomena are not fully understood. However, it is clear that the prime source of auroral particles is the solar wind feeding the magnetosphere, the reservoir containing the radiation zones and temporarily magnetically-trapped particles confined by the geomagnetic field, coupled with particle acceleration processes.

(Excerpt end)

Observation:

Cosmologists admit the aurora "phenomena are not fully understood." This is because the solar wind is invisible and the acceptance of that cause is difficult when gravity and visible mass is often expected as a cause.

A web site "everything electric" has a page titled Birkelund Currents

Excerpt:

More than a century ago, Norwegian scientist Kristian Birkeland proposed that vast electric currents powered by solar wind were travelling through Earth's ionosphere by the planet's magnetic field ...
Known as Birkeland currents, they carry up to 1 TW of electric power to the upper atmosphere - about a third of the total power consumption of the US in a year. They're also responsible for the aurora borealis and aurora australis that light up the poles of the Northern and Southern Hemispheres ...
satellites detected incredibly large electrical fields, which are generated in the ionosphere where upwards and downwards Birkeland currents are interacting above the planet ... the satellite trio has discovered what these electrical fields are driving - extreme supersonic plasma jets that have been dubbed 'Birkeland current boundary flows'.

(Excerpt end)

This page included this image portraying these currents:

# Birkeland Currents

The correct spelling is Birkelund, named after Norwegian scientist Kristian Birkeland who is best known for explaining the natural phenomenon of Aurora Borealis in great detail.

## 17.4.1 Moon

Moon's far side round craters.

China took many images of the Moon's far side but they did not generate much publicity.

There are some noticeable differences to the near side.

Background information for the first image:

Tycho is the well known, bright crater on the Moon with distinct rays and a central peak on the flat floor with roughly vertical walls around the circumference.


As Ralph Juergens suggested in the 1970's, Tycho was electrically machined by an electrical current pair which, while twisting, carved the flat floor and the walls with a consistent radius but the gap between the filaments left the central peak.

The initial discharge resulted in the rays when surface electrons were drawn to the discharge point. The rays from craters are surface electrical scarring, not ejected material which must result from a supposed impact explosion.

On the far side in this image, there is no dominant crater like Tycho on the near side.
Instead there are many round craters with flat floors and no central peak and no rays. I will call this a frequent Flat Floor Crater, or just FFC.

Roughly to the right of center is a vertical line of 4 FFC. The top 1 of the 4 has 2 smaller FFC in a line to its lower right.

The second from the top also has 2 smaller FFC to its lower right, but these 3 FFC have no space between rims.

Above and to the right is a large round crater with a flat floor and a peak, but no rays.
Below that crater, to the right of the vertical 4, is a vertical line of 6 FFC.

At the far top right there is a line of about 5 FFC.

At the far bottom left is a bright crater with 2 associated rays. Despite the rays this crater lacks the depth of Tycho. Near the bottom is a roughly horizontal line of 6 FFC.

Observation:

This far side image reveals many craters whose electrical discharge machined a circle but not to a substantial depth.

Roughly in the center of the image are 3 flat plains with no prominent craters. However the bottom plain has associated features including FFC on its right edge and on its lower left edge. To its left is a small rayed crater which has some depth but not enough depth for a flat floor.

With this image resolution we are missing the many small craters seen on the near side having a sharp circular rim.

First image of far side craters in References

A second image is much the same as the first, for the curious. The one noticeable difference is the dark area near the top. This does not look like the flat plains on the near side which are not this dark.

One can only wonder when more far side images will be available.

From NASA: The Clementine mission mapped most of the lunar surface at a number of resolutions and wavelengths from UV to IR.

However, from Wikipedia:

Clementine (officially called the Deep Space Program Science Experiment (DSPSE)) was a joint space project between the Ballistic Missile Defense Organization (BMDO, previously the Strategic Defense Initiative Organization, or SDIO) and NASA.

We probably won't see all the images from Clementine.

## 17.5 Planet Mars

NASA had a story about the Mars aurora, titled

NASA's MAVEN Spacecraft Finds That "Stolen" Electrons Enable Unusual Aurora on Mars

Excerpt:

Auroras flare up when energetic particles plunge into a planet's atmosphere, bombarding gases and making them glow. While electrons generally cause this natural phenomenon, sometime protons can elicit the same response, although it's more rare. Now, the MAVEN team has learned that protons were doing at Mars the same thing as electrons usually do at Earth—create aurora. This is especially true when the Sun ejects a particularly strong pulse of protons, which are hydrogen atoms stripped of their lone electrons by intense heat. The Sun ejects protons at speeds up to two million miles per hour (more than 3 million kilometers per hour) in an erratic flow called the solar wind.

(Excerpt end)

Observation:

Cosmologists often avoid the word proton and instead use the term hydrogen ion. NASA used proton.
A hydrogen atom loses its electron when it absorbs the energy in electromagnetic radiation above the Lyman limit.

Atoms do not carry heat or temperature; they carry kinetic energy in their motion.

## 17.5.1 Determining the age of surfaces on Mars

That is the name of a web page.

This is a study of the Martian surface and the distribution of craters. The conclusion has 3 defined periods of bombardment for Mars and so the surface of Mars has 3 regions.
The article has a link to a discussion of crater forms, with this excerpt:

Certainly the ejecta resembles patterns that one might see throwing rocks into fluid mud, and for this reason these craters are often called "splosh" or "fluidized ejecta" craters. However, one must be careful not to predict the behavior of impactors moving at many kilometers per second from pebbles tossed into mud, and the question of exactly what factors contribute to the form of these ejecta patterns must wait for the further exploration of the planet.

Some [craters] do resemble those seen on the Moon - for example, those smaller than 5 km in diameter are usually bowl-shaped, with raised rims and slightly flat floors, just like similarly sized craters on the Moon.

In general, the more craters appear on a surface, the older that surface is. But like most principles in the real world, that rule must be applied with caution.

(Excerpt end)

Observation:

The previous book Cosmology Transition described the many round craters in the solar system. Most round craters appear electrical in nature, not explosions upon impact. This description of Martian craters brings in water under the surface to explain the varying terrain around the craters. No water is involved in forming lunar craters. There are similar crater types on Mars and Moon.
One must proceed with "caution" when offering explanations with water while "[waiting] for the further exploration of the planet."

## 17.5.2 Valles Marineris

There are articles about the large canyon on Mars. Here is general information.

Excerpt from Wikipedia:

Valles Marineris is a system of canyons that runs along the Martian surface east of the Tharsis region. At more than 4,000 km (2,500 mi) long, 200 km (120 mi) wide and up to 7 km (23,000 ft) deep, Valles Marineris is one of the largest canyons of the Solar System, surpassed in length only by the rift valleys of Earth.

There have been many different theories about the formation of Valles Marineris that have changed over the years. Ideas in the 1970s were erosion by water or thermokarst activity, which is the melting of permafrost in glacial climes. Thermokarst activity may contribute, but erosion by water is a problematic mechanism because liquid water cannot exist in most current Martian surface conditions.

(Excerpt end)

Observation:

Currently every theory for the canyon's formation is "problematic."

Thunderbolts site had several articles on Mars including its surface and its large canyon.

One was titled:

## The Thunderbolt that Changed the Face of Mars

Excerpt:

As in arc welding, material from the electrode will be accelerated upwards against gravity. This means that any electrical event capable of creating the Valles Marineris on Mars would likely throw huge volumes of rock into space, creating vast debris clouds. Some of the ejecta would encircle Mars or fall back to the planet, while other material would presumably escape the battlefield altogether, to be encountered by the Earth and other planets across the millennia.
The electric interpretation thus removes another conundrum. It explains why Martian meteorites have arrived at Earth to perplex physicists and geologists. Until the signature of Martian atmosphere in these meteorites was identified beyond any reasonable doubt, the experts said that such rocks could not achieve escape velocity from Mars without being vaporized by the explosive force required. Electrical acceleration, however, faces no such dilemma.
For millennia afterwards we might also expect the Earth to periodically encounter clouds of rusty red dust, the residue of Martian surface material removed by interplanetary lightning. This too appears to have occurred right up to the twentieth century.
The event that created Valles Marineris lofted into space something like 10,000 trillion tons of rock and dust. This can only mean that there have been far more Martian meteorites and falls of dust on Earth than geologists have recognized.

(Excerpt end)

Observation:

This single electrical explanation for the canyon solves the problem of meteorites on Earth known to have content consistent with the Martian surface.

## 17.5.3 Dust Devils

Thunderbolts site had an article on these Dust Devils on Earth, which are similar to those on Mars, titled:
Electric Devils


Excerpt:

In an [electric cosmology], no collisions from bouncing sand grains are necessary. Charge separation already exists in the atmosphere. Without clouds like those on Earth to send lightning down to ground level, the electric discharges on Mars form giant whirlwinds that are part of an interplanetary electric circuit. It is the kind of circuit that drives weather systems on Earth. If this is true, then Martian "dust devils" and those on Earth are both illustrations of how electricity behaves in the solar system.

(Excerpt end)

Observation:
Despite differences in atmospheric composition and density, the electrical mechanism is the same on Earth and Mars.

## 17.6 Planet Jupiter

Excerpt from Wikipedia:

Jupiter's magnetic field is fourteen times as strong as that of Earth, ranging from 4.2 gauss (0.42 mT) at the equator to 10–14 gauss (1.0–1.4 mT) at the poles, making it the strongest in the Solar System (except for sunspots). This field is thought to be generated by eddy currents—swirling movements of conducting materials—within the liquid metallic hydrogen core. The volcanoes on the moon Io emit large amounts of sulfur dioxide forming a gas torus along the moon's orbit. The gas is ionized in the magnetosphere producing sulfur and oxygen ions. They, together with hydrogen ions originating from the atmosphere of Jupiter, form a plasma sheet in Jupiter's equatorial plane. The plasma in the sheet co-rotates with the planet causing deformation of the dipole magnetic field into that of magnetodisk. Electrons within the plasma sheet generate a strong radio signature that produces bursts in the range of 0.6–30 MHz.
At about 75 Jupiter radii from the planet, the interaction of the magnetosphere with the solar wind generates a bow shock. Surrounding Jupiter's magnetosphere is a magnetopause, located at the inner edge of a magnetosheath—a region between it and the bow shock. The solar wind interacts with these regions, elongating the magnetosphere on Jupiter's lee side and extending it outward until it nearly reaches the orbit of Saturn. The four largest moons of Jupiter all orbit within the magnetosphere, which protects them from the solar wind.

The magnetosphere of Jupiter is responsible for intense episodes of radio emission from the planet's polar regions. Volcanic activity on Jupiter's moon Io injects gas into Jupiter's magnetosphere, producing a torus of particles about the planet. As Io moves through this torus, the interaction generates Alfvén waves that carry ionized matter into the polar regions of Jupiter. As a result, radio waves are generated through a cyclotron maser mechanism, and the energy is transmitted out along a cone-shaped surface. When Earth intersects this cone, the radio emissions from Jupiter can exceed the solar radio output.

(Excerpt end)

Observation:

The strength of its magnetic field requires a very strong electric current.
The radio emissions are low energy synchrotron radiation, so there must be charged particles in motion. A "cyclotron" is mentioned which is another form of a synchrotron.

NASA had a story titled:

Hubble Captures Vivid Auroras in Jupiter's Atmosphere

Excerpt:

"These auroras are very dramatic and among the most active I have ever seen", said Jonathan Nichols from the University of Leicester, U.K., and principal investigator of the study. "It almost seems as if Jupiter is throwing a firework party for the imminent arrival of Juno."

To highlight changes in the auroras Hubble is observing Jupiter almost daily for several months. Using this series of far-ultraviolet images from Hubble's Space Telescope Imaging Spectrograph, it is possible for scientists to create videos that demonstrate the movement of the vivid auroras, which cover areas bigger than Earth.

Not only are the auroras huge in size, they are also hundreds of times more energetic than auroras on Earth. And, unlike those on Earth, they never cease. While on Earth the most intense auroras are caused by solar storms — when charged particles rain down on the upper atmosphere, excite gases and cause them to glow red, green and purple — Jupiter has an additional source for its auroras.

The strong magnetic field of the gas giant grabs charged particles from its surroundings. This includes not only the charged particles within the solar wind but also the particles thrown into space by its orbiting moon Io, known for its numerous and large volcanoes.

The new observations and measurements made with Hubble and Juno will help to better understand how the sun and other sources influence auroras. While the observations with Hubble are still ongoing and the analysis of the data will take several more months, the first images and videos are already available and show the auroras on Jupiter's north pole in their full beauty.

(Excerpt end)

Observation:

Jupiter with its moons creates its own electrical system, gathering charged particles (plasma) from Sun and even from Io, creating an aurora. The 4 largest moons orbit within the magnetosphere.

## 17.6.1 Ganymede

Ganymede is Jupiter's largest moon and is also the largest moon in the solar system.

Excerpt from Wikipedia:

The Galileo craft made six close flybys of Ganymede from 1995–2000 and discovered that Ganymede has a permanent (intrinsic) magnetic moment independent of the Jovian magnetic field. The value of the moment is about $1.3 \times 10^{13}$ T·m3, which is three times larger than the magnetic moment of Mercury. The magnetic dipole is tilted with respect to the rotational axis of Ganymede by 176°, which means that it is directed against the Jovian magnetic moment. Its north pole lies below the orbital plane. The dipole magnetic field created by this permanent moment has strength of $719 \pm 2$ nT at Ganymede's equator, which should be compared with the Jovian magnetic field at the distance of Ganymede—about 120 nT. The equatorial field of Ganymede is directed against the Jovian field, meaning reconnection is possible. The intrinsic field strength at the poles is two times that at the equator [is] 1440 nT.

The permanent magnetic moment carves a part of space around Ganymede, creating a
tiny magnetosphere embedded inside that of Jupiter; it is the only moon in the Solar System known to possess the feature. Its diameter is 4–5 Ganymede radii. The Ganymedian magnetosphere has a region of closed field lines located below 30° latitude, where charged particles (electrons and ions) are trapped, creating a kind of radiation belt.

The main ion species in the magnetosphere is single ionized oxygen—O+ - which fits well with Ganymede's tenuous oxygen atmosphere. In the polar cap regions, at latitudes higher than 30°, magnetic field lines are open, connecting Ganymede with Jupiter's ionosphere. In these areas, the energetic (tens and hundreds of kiloelectronvolt) electrons and ions have been detected, which may cause the auroras observed around the Ganymedian poles. In addition, heavy ions precipitate continuously on Ganymede's polar surface, sputtering and darkening the ice. The interaction between the Ganymedian magnetosphere and Jovian plasma is in many respects similar to that of the solar wind and Earth's magnetosphere. The plasma co-rotating with Jupiter impinges on the trailing side of the Ganymedian magnetosphere much like the solar wind impinges on the Earth's magnetosphere. The main difference is the speed of plasma flow—supersonic in the case of Earth and subsonic in the case of Ganymede. Because of the subsonic flow, there is no bow shock off the trailing hemisphere of Ganymede.

In addition to the intrinsic magnetic moment, Ganymede has an induced dipole magnetic field. Its existence is connected with the variation of the Jovian magnetic field near Ganymede. The induced moment is directed radially to or from Jupiter following the direction of the varying part of the planetary magnetic field. The induced magnetic moment is an order of magnitude weaker than the intrinsic one.

The field strength of the induced field at the magnetic equator is about 60 nT—half of that of the ambient Jovian field. The induced magnetic field of Ganymede is similar to those of Callisto and Europa, indicating that Ganymede also has a subsurface water ocean with a high electrical conductivity.

Given that Ganymede is completely differentiated and has a metallic core, its intrinsic magnetic field is probably generated in a similar fashion to the Earth's: as a result of conducting material moving in the interior. The magnetic field detected around Ganymede is likely to be caused by compositional convection in the core, if the magnetic field is the product of dynamo action, or magnetoconvection. Despite the presence of an iron core, Ganymede's magnetosphere remains enigmatic, particularly given that similar bodies lack the feature. Some research has suggested that, given its relatively small size, the core ought to have sufficiently cooled to the point where fluid motions, hence a magnetic field would not be sustained. One explanation is that the same orbital resonances proposed to have disrupted the surface also allowed the magnetic field to persist: with Ganymede's eccentricity pumped and tidal heating of the mantle increased during such resonances, reducing heat flow from the core, leaving it fluid and convective. Another explanation is a remnant magnetization of silicate rocks in the mantle, which is possible if the satellite had a more significant dynamo-generated field in the past.

Voyager 1 and Voyager 2 were passing by Ganymede in 1979. They refined its size, revealing it was larger than Saturn's moon Titan, which was previously thought to have been bigger. The grooved terrain was also seen.

In 1995, the Galileo spacecraft entered orbit around Jupiter and between 1996 and 2000 made six close flybys to explore Ganymede. During the closest flyby, Galileo passed just 264 km from the surface of Ganymede. During a flyby in 1996, the Ganymedian magnetic field was discovered, while the discovery of the ocean was announced in 2001. Galileo transmitted a large number of spectral images and discovered several non-ice compounds on the surface of Ganymede.

The most recent close observations of Ganymede were made by New Horizons, which recorded topographic and compositional mapping data of Europa and Ganymede during its flyby of Jupiter in 2007 en route to Pluto.
On December 25, 2019, the Juno spacecraft flew by Ganymede and got images of the moon's polar regions. The images were taken at a range of 97,680 to 109,439 kilometers (60,696 to 68,002 mi).

(Excerpt end)

Observation:

The statement "reconnection is possible" is wrong because magnetic reconnection is a proposed event which does not exist. Field lines are an abstraction to represent field strength. They do not exist, just like lines of latitude do not exist. The "magnetosphere remains enigmatic" because the known mechanism for generating of a magnetic field by an electric current is not being applied,

Ganymede has a tenuous atmosphere, water, and a magnetic field for protection from cosmic rays at the surface. Others have remarked this is a moon worth further investigation.

## 17.7 Planet Saturn

Saturn is the second largest planet; Jupiter is first.

Excerpt from Wikipedia:

Saturn's interior is most likely composed of a core of iron–nickel and rock (silicon and oxygen compounds). Its core is surrounded by a deep layer of metallic hydrogen, an intermediate layer of liquid hydrogen and liquid helium, and finally a gaseous outer layer. Saturn has a pale yellow hue due to ammonia crystals in its upper atmosphere.
An electrical current within the metallic hydrogen layer is thought to give rise to Saturn's planetary magnetic field, which is weaker than the Earth's, but has a magnetic moment 580 times that of Earth due to Saturn's larger size. Saturn's magnetic field strength is around one-twentieth of Jupiter's. The outer atmosphere is generally bland and lacking in contrast, although long-lived features can appear. Wind speeds on Saturn can reach 1,800 km/h (1,100 mph; 500 m/s), higher than on Jupiter, but not as high as those on Neptune.
In January 2019, astronomers reported that a day on the planet Saturn has been determined to be 10h 33m 38s, based on studies of the planet's C Ring.

(Excerpt end)

The story title about that 2019 rotation study:

Making sense of Saturn's impossible rotation

Excerpt:

Saturn may be doing a little electromagnetic shimmy and twist which has been throwing off attempts by scientists to determine how long it takes for the planet to rotate on its axis, according to a new study.

Saturn emits only low-frequency radio patterns that are blocked by Earth's atmosphere, making it difficult to study Saturn's rotation from the Earth's surface. In contrast, Jupiter emits radio patterns at higher frequencies that allowed radio astronomers to work out its rotation period before the space age got well under way.

It wasn't until spacecraft were sent to Saturn that scientists were able to collect data about its rotation. Voyagers 1 and 2 sent home the first hints of Saturn's rotation in 1980 and 1981. They detected a modulation of radio intensity that suggested the planet rotated once every 10 hours and 40 minutes.

"So that was what was called the rotation period," said Duane Pontius of Birmingham-Southern College in Alabama and a co-author of the new study.

When the Cassini spacecraft arrived at Saturn 23 years later to study the planet for 13 years, it found something astonishing. "In about 2004 we saw the period had changed by 6 minutes, about 1 percent," Pontius said.

But how does an entire planet change the speed of its rotation in 20 years? That's the sort of change that takes hundreds of millions of years. Even more mysterious was Cassini's detection of electromagnetic patterns that suggested the planet's rotation is different in the northern and southern hemispheres.

"For a long time, I assumed there was something wrong with the data interpretation," Pontius recalled. "It's just not possible."

To find out what was really going on, Pontius and his co-authors started by looking at how Saturn is different from its closest sibling, Jupiter.

According to the model being proposed by Pontius and his colleagues, the variations in UV from summer to winter in the different hemispheres affects the plasma so that it creates more or less drag at the altitudes where it encounters the planet's gaseous atmosphere.
That difference in drag makes the atmosphere slow down, which is what sets the period seen in the radio signals. Change the plasma seasonally, and you change the period of the radio emissions, which is what is seen on Saturn.
The new model provides a solution to the puzzle of Saturn's impossible changing rotation periods. It also shows that the observed periods are not the rotation period of Saturn's core, which remains unmeasured.
Pontius presented the model earlier this year at a meeting of Saturn scientists and said it was well received. Now he hopes that other researchers will take the next step to refine the model by exploring how well it fits with 13 years of Saturn data collected by Cassini.

(Excerpt end)

Observation:

The explanation for the change in rotation is essentially "drag."

Perhaps, the solar rotation is relevant. Our Sun has different rotation periods for different latitudes. The explanation for that solar inconsistency is not drag. According to Wikipedia: "The source of this differential rotation is an area of current research in solar astronomy."

As noted in the second book Cosmology Transition, both electric cosmology and plasma cosmology are proposing a solar model based on liquid metallic hydrogen. The Sun is not a sphere of gas. Both Jupiter and Saturn have models based on liquid metallic hydrogen for a substantial portion of their mass. Perhaps cosmology will find explanations when recognizing such similarities.

As noted earlier in this book, the only mechanism for radiating the radio frequency range is synchrotron radiation. When an electric current bends its path by a magnetic field, like in a synchrotron, synchrotron radiation results. The velocity of the electrons or the current determines the maximum frequency. X-ray machines use this method. The radio frequency is at the low end of a synchrotron.

Saturn's radio emissions varying with the planet's rotation are driven by slow moving electrons affected by a magnetic field. Saturnian kilometric radiation is mentioned below with the Saturn aurora.

## 17.7.1 Magnetosphere of Saturn

Excerpt from Wikipedia:

The magnetosphere of Saturn is the cavity created in the flow of the solar wind by the planet's internally generated magnetic field.
Saturn's magnetosphere is the second largest of any planet in the Solar System after Jupiter.
The magnetopause, the boundary between Saturn's magnetosphere and the solar wind, is located at a distance of about 20 Saturn radii from the planet's center, while its magnetotail stretches hundreds of Saturn radii behind it.
Saturn's magnetosphere is filled with plasmas originating from both the planet and its moons. The main source is the small moon Enceladus, which ejects as much as 1,000 kg/s of water vapor from the geysers on its south pole, a portion of which is ionized and forced to co-rotate with the Saturn's magnetic field. This loads the field with as much as 100 kg of water group ions per second. This plasma gradually moves out from the inner magnetosphere via the interchange instability mechanism and then escapes through the magnetotail.
The interaction between Saturn's magnetosphere and the solar wind generates bright oval aurorae around the planet's poles observed in visible, infrared and ultraviolet light. The aurorae are related to the powerful saturnian kilometric radiation (SKR), which spans the frequency interval between 100 kHz to 1300 kHz and was once thought to modulate with a period equal to the planet's rotation. However, later measurements showed that the periodicity of the SKR's modulation varies by as much as 1%, and so probably does not exactly coincide with Saturn's true rotational period, which as of 2010 remains unknown.

Inside the magnetosphere there are radiation belts, which house particles with energy as high as tens of megaelectronvolts. The energetic particles have significant influence on the surfaces of inner icy moons of Saturn.

(Excerpt end)
Observation:
There is an "internally generated magnetic field" which needs an explanation.
Thunderbolts has a story titled
Saturn Supernova

Excerpt:

Electrical theorists argue that Saturn moves within the plasmasphere of the Sun and interacts with the Sun's electric field. Planets and moons in the Solar System are charged bodies. They are not isolated in "empty" space, but "converse" electrically with each other. Enceladus, Dione and Tethys all move within the plasmasphere of Saturn, so it is only to be expected that they would transact electrically with their primary. The simplest, most straightforward explanation for the charged particle acceleration is electric discharge, so there is no need to conjure implausible internal dynamics to account for them.
In fact, other investigators admit to Saturn's interactions with its moons: "…we conclude that the observed double-peaked ('butterfly') pitch-angle distributions [of the plasma] result from the transport of plasma from regions near the orbits of Dione and Tethys, supporting the idea of distinct plasma tori associated with these moons."

Saturn emits 2.3 times more energy than it receives from the Sun. 90 megawatts of X-rays were detected by Cassini. But even that was not attributed to its electrical nature. Instead, Saturn's atmosphere is said to reflect X-rays from the Sun, although the science team admitted when the discovery was made that the intensity of the "reflections" was "surprising." The reason it was so surprising is that they ignored the fact that planets with magnetic fields can capture ionized particles to form
a giant electrified magnetosphere.
It is that magnetosphere that traps and accelerates the charged particles from Saturn.

(Excerpt end)

Observation:

Plasma behaviors are ignored by popular cosmology.

## 17.7.2 Titan

Excerpt from Wikipedia:

Titan is the largest moon of Saturn and the second-largest natural satellite in the Solar System. It is the only moon known to have a dense atmosphere, and the only known body in space, other than Earth, where clear evidence of stable bodies of surface liquid has been found.

The Cassini probe discovered the evidence for the layered structure in the form of natural extremely-low-frequency radio waves in Titan's atmosphere. Titan's surface is thought to be a poor reflector of extremely-low-frequency radio waves, so they may instead be reflecting off the liquid–ice boundary of a subsurface ocean. Surface features were observed by the Cassini spacecraft to systematically shift by up to 30 kilometers (19 mi) between October 2005 and May 2007, which suggests that the crust is decoupled from the interior, and provides additional evidence for an interior liquid layer. Further supporting evidence for a liquid layer and ice shell decoupled from the solid core comes from the way the gravity field varies as Titan orbits Saturn. Comparison of the gravity field with the RADAR-based topography observations also suggests that the ice shell may be substantially rigid.

Observations of it made in 2004 by Cassini suggest that Titan is a "super rotator", like Venus, with an atmosphere that rotates much faster than its surface. Observations from the Voyager space probes have shown that Titan's atmosphere is denser than Earth's, with a surface pressure about 1.45 atm. It is also about 1.19 times as massive as Earth's overall, or about 7.3 times more massive on a per surface area basis. Opaque haze layers block most visible light from the Sun and other sources and obscure Titan's surface features. Titan's lower gravity means that its atmosphere is far more extended than Earth's. The atmosphere of Titan is opaque at many wavelengths and as a result, a complete reflectance spectrum of the surface is impossible to acquire from orbit. It was not until the arrival of the Cassini–Huygens spacecraft in 2004 that the first direct images of Titan's surface were obtained.

(Excerpt end)

Observation:

There are two notable features among the others:

a) "natural extremely-low-frequency radio waves in Titan's atmosphere" means there is a mechanism for synchrotron radiation between the internal layers.

b) "super rotator" means there is an unusual wind mechanism here. Given the lack of solar heating disparities to drive wind in this opaque atmosphere, there must be an electrical mechanism for the wind velocity, which was also suggested by the "natural" radio emissions.

## 17.8 Planet Uranus

Excerpt from Wikipedia:

The standard model of Uranus' structure is that it consists of three layers: a rocky (silicate/iron–nickel) core in the centre, an icy mantle in the middle and an outer gaseous hydrogen/helium envelope. The core is relatively small, with a mass of only 0.55 Earth masses and a radius less than 20% of Uranus'; the mantle comprises its bulk, with around 13.4 Earth masses, and the upper atmosphere is relatively insubstantial, weighing about 0.5 Earth masses and extending for the last 20% of Uranus' radius. Uranus' core density is around 9 g/cm$^3$, with a pressure in the centre of 8 million bars (800 GPa) and a temperature of about 5000 K. The ice mantle is not in fact composed of ice in the conventional sense, but of a hot and dense fluid consisting of water, ammonia and other volatiles. This fluid, which has a high electrical conductivity, is sometimes called a water–ammonia ocean.
The extreme pressure and temperature deep within Uranus may break up the methane molecules, with the carbon atoms condensing into crystals of diamond that rain down through the mantle like hailstones. Very-high-pressure experiments at the Lawrence Livermore National Laboratory suggest that the base of the mantle may comprise an ocean of liquid diamond, with floating solid 'diamond-bergs'. Scientists also believe that rainfalls of solid diamonds occur on Uranus, as well as on Jupiter, Saturn, and Neptune.

The bulk compositions of Uranus and Neptune are different from those of Jupiter and Saturn, with ice dominating over gases, hence justifying their separate classification as ice giants. There may be a layer of ionic water where the water molecules break down into a soup of hydrogen and oxygen ions, and deeper down superionic water in which the oxygen crystallises but the hydrogen ions move freely within the oxygen lattice.

Although the model considered above is reasonably standard, it is not unique; other models also satisfy observations. For instance, if substantial amounts of hydrogen and rocky material are mixed in the ice mantle, the total mass of ices in the interior will be lower, and, correspondingly, the total mass of rocks and hydrogen will be higher. Presently available data does not allow a scientific determination of which model is correct. The fluid interior structure of Uranus means that it has no solid surface. The gaseous atmosphere gradually transitions into the internal liquid layers. For the sake of convenience, a revolving oblate spheroid set at the point at which atmospheric pressure equals 1 bar (100 kPa) is conditionally designated as a "surface". It has equatorial and polar radii of 25,559 ± 4 km (15,881.6 ± 2.5 mi) and 24,973 ± 20 km (15,518 ± 12 mi), respectively. This surface is used throughout this article as a zero point for altitudes.

Voyager 2's observations [in 1986] revealed that Uranus' magnetic field is peculiar, both because it does not originate from its geometric centre, and because it is tilted at 59° from the axis of rotation. In fact the magnetic dipole is shifted from Uranus' centre towards the south rotational pole by as much as one third of the planetary radius.

This unusual geometry results in a highly asymmetric magnetosphere, where the magnetic field strength on the surface in the southern hemisphere can be as low as 0.1 gauss (10 µT), whereas in the northern hemisphere it can be as high as 1.1 gauss (110 µT). The average field at the surface is 0.23 gauss (23 µT). Studies of Voyager 2 data in 2017 suggest that this asymmetry causes Uranus' magnetosphere to connect with the solar wind once a Uranian day, opening the planet to the Sun's particles. In comparison, the magnetic field of Earth is roughly as strong at either pole, and its "magnetic equator" is roughly parallel with its geographical equator. The dipole moment of Uranus is 50 times that of Earth. Neptune has a similarly displaced and tilted magnetic field, suggesting that this may be a common feature of ice giants. One hypothesis is that, unlike the magnetic fields of the terrestrial and gas giants, which are generated within their cores, the ice giants' magnetic fields are generated by motion at relatively shallow depths, for instance, in the water–ammonia ocean. Another possible explanation for the magnetosphere's alignment is that there are oceans of liquid diamond in Uranus' interior that would deter the magnetic field.

Despite its curious alignment, in other respects the Uranian magnetosphere is like those of other planets: it has a bow shock at about 23 Uranian radii ahead of it, a magnetopause at 18 Uranian radii, a fully developed magnetotail, and radiation belts. Overall, the structure of Uranus' magnetosphere is different from Jupiter's and more similar to Saturn's. Uranus' magnetotail trails behind it into space for millions of kilometres and is twisted by its sideways rotation into a long corkscrew.

Uranus' magnetosphere contains charged particles: mainly protons and electrons, with a small amount of H2+ ions. Many of these particles probably derive from the thermosphere. The ion and electron energies can be as high as 4 and 1.2 megaelectronvolts, respectively. The density of low-energy (below 1 kiloelectronvolt) ions in the inner magnetosphere is about 2 $cm^{-3}$. The particle population is strongly affected by the Uranian moons, which sweep through the magnetosphere, leaving noticeable gaps. The particle flux is high enough to cause darkening or space weathering of their surfaces on an astronomically rapid timescale of 100,000 years. This may be the cause of the uniformly dark colouration of the Uranian satellites and rings. Uranus has relatively well developed aurorae, which are seen as bright arcs around both magnetic poles. Unlike Jupiter's, Uranus' aurorae seem to be insignificant for the energy balance of the planetary thermosphere.

In March 2020, NASA astronomers reported the detection of a large atmospheric magnetic bubble, also known as a plasmoid, released into outer space from the planet Uranus, after reevaluating old data recorded by the Voyager 2 space probe during a flyby of the planet in 1986.

(Excerpt end)

Observation:

[The mantle] fluid has a high electrical conductivity."

That suggests internal electric currents can drive the measured magnetic fields above them.

This quote from above is noteworthy:
"Presently available data does not allow a scientific determination of which model is correct. "

Cosmologists are still working on a model for this planet's internal structure.

Just as cosmology often misses electromagnetic and plasma behaviors, planetary magnetic fields remain a mystery.

### 17.8.1 Uranus Rings

ALMA imaged the rings around planet Uranus.

One interesting observation:

"The rings of Uranus are compositionally different from Saturn's main ring, in the sense that in optical and infrared, the albedo is much lower: they are really dark, like charcoal,"

Rings could be debris remaining from the time of formation 4.5 billion years ago.

(Excerpt end)

Observation:

This is the diversity of rings:

Saturn's rings are fragments of mostly water ice with a trace component of rock.

Neptune's rings are dark like the rings of Uranus except Neptune's outer Adams ring has bright arcs.

Jupiter's rings are dust.

With different compositions for each planet, perhaps each formed from a different source or mechanism.

## 17.9 Planet Neptune

Uranus and Neptune are not as similar as one might expect for the two most distant giant planets.

Excerpt from Wikipedia:

Like Jupiter and Saturn, Neptune's atmosphere is composed primarily of hydrogen and helium, along with traces of hydrocarbons and possibly nitrogen, though it contains a higher proportion of "ices" such as water, ammonia and methane. However, similar to Uranus, its interior is primarily composed of ices and rock; Uranus and Neptune are normally considered "ice giants" to emphasise this distinction. Traces of methane in the outermost regions in part account for the planet's blue appearance.

In contrast to the hazy, relatively featureless atmosphere of Uranus, Neptune's atmosphere has active and visible weather patterns. For example, at the time of the Voyager 2 flyby in 1989, the planet's southern hemisphere had a Great Dark Spot comparable to the Great Red Spot on Jupiter. These weather patterns are driven by the strongest sustained winds of any planet in the Solar System, with recorded wind speeds as high as 2,100 km/h (580 m/s; 1,300 mph). Because of its great distance from the Sun, Neptune's outer atmosphere is one of the coldest places in the Solar System, with temperatures at its cloud tops approaching 55 K (−218 °C; −361 °F). Temperatures at the planet's centre are approximately 5,400 K (5,100 °C; 9,300 °F). Neptune has a faint and fragmented ring system (labelled "arcs"), which was discovered in 1984, then later confirmed by Voyager 2.

(Excerpt end)

Observations:

Neptune has the strongest winds in the solar system but its atmosphere is the furthest from the Sun. This suggests the winds are not driven by uneven heating from the weak solar radiation but rather an electrical mechanism. This was also noted with Uranus whose winds are in "super" rotation.

Neptune is even more complicated at its greater distance.

Excerpt continued:

Unlike Uranus, Neptune's composition has a higher volume of ocean, whereas Uranus has a smaller mantle. Neptune's spectra suggest that its lower stratosphere is hazy due to condensation of products of ultraviolet photolysis of methane, such as ethane and ethyne. The stratosphere is also home to trace amounts of carbon monoxide and hydrogen cyanide. The stratosphere of Neptune is warmer than that of Uranus due to the elevated concentration of hydrocarbons. For reasons that remain obscure, the planet's thermosphere is at an anomalously high temperature of about 750 K. The planet is too far from the Sun for this heat to be generated by ultraviolet radiation. One candidate for a heating mechanism is atmospheric interaction with ions in the planet's magnetic field. Other candidates are gravity waves from the interior that dissipate in the atmosphere. The thermosphere contains traces of carbon dioxide and water, which may have been deposited from external sources such as meteorites and dust.

Neptune's more varied weather when compared to Uranus is due in part to its higher internal heating.

The upper regions of Neptune's troposphere reach a low temperature of 51.8 K (−221.3 °C). At a depth where the atmospheric pressure equals 1 bar (100 kPa), the temperature is 72.00 K (−201.15 °C). Deeper inside the layers of gas, the temperature rises steadily. As with Uranus, the source of this heating is unknown, but the discrepancy is larger: Uranus only radiates 1.1 times as much energy as it receives from the Sun; whereas Neptune radiates about 2.61 times as much energy as it receives from the Sun. Neptune is the farthest planet from the Sun, and lies over 50% farther from the Sun than Uranus, and receives only 40% its amount of sunlight, yet its internal energy is sufficient to drive the fastest planetary winds seen in the Solar System. Depending on the thermal properties of its interior, the heat left over from Neptune's formation may be sufficient to explain its current heat flow, though it is more difficult to simultaneously explain Uranus's lack of internal heat while preserving the apparent similarity between the two planets.

Neptune resembles Uranus in its magnetosphere, with a magnetic field strongly tilted relative to its rotational axis at 47° and offset at least 0.55 radii, or about 13,500 km from the planet's physical centre. Before Voyager 2's arrival at Neptune, it was hypothesised that Uranus's tilted magnetosphere was the result of its sideways rotation. In comparing the magnetic fields of the two planets, scientists now think the extreme orientation may be characteristic of flows in the planets' interiors. This field may be generated by convective fluid motions in a thin spherical shell of electrically conducting liquids (probably a combination of ammonia, methane and water) resulting in a dynamo action. The dipole component of the magnetic field at the magnetic equator of Neptune is about 14 microteslas (0.14 G).

By contrast, Earth, Jupiter and Saturn have only relatively small quadrupole moments, and their fields are less tilted from the polar axis. The large quadrupole moment of Neptune may be the result of offset from the planet's centre and geometrical constraints of the field's dynamo generator. Neptune's bow shock, where the magnetosphere begins to slow the solar wind, occurs at a distance of 34.9 times the radius of the planet. The magnetopause, where the pressure of the magnetosphere counterbalances the solar wind, lies at a distance of 23–26.5 times the radius of Neptune. The tail of the magnetosphere extends out to at least 72 times the radius of Neptune, and likely much farther.

(Excerpt end)

Observation:

The measured temperatures remain "for reasons that remain obscure."
It is admitted "the heat left over from Neptune's formation may be sufficient to explain [some but not all features]."

That assumes there is no other energy source than sunlight to this distant planet. Other planets, like the Earth appear to have an electric current from the Sun. That possible external energy source for Neptune must be considered.

Like with Uranus, Just as cosmology often misses electromagnetic and plasma behaviors, the source of energy for some features of Neptune remains a mystery. It must be electrical.

## 17.9.1 Triton

Excerpt from Wikipedia:

During its 1989 flyby of Triton, Voyager 2 found surface temperatures of 38 K (−235 °C) and also discovered active geysers; Voyager 2 remains the only spacecraft to visit Triton. Triton is one of the few moons in the Solar System known to be geologically active (the others being Jupiter's Io and Europa, and Saturn's Enceladus and Titan). As a consequence, its surface is relatively young, with few obvious impact craters. Intricate cryovolcanic and tectonic terrains suggest a complex geological history. Part of its surface has geysers erupting sublimated nitrogen gas, contributing to a tenuous nitrogen atmosphere less than 1/70,000 the pressure of Earth's atmosphere at sea level.
Triton's mean density implies that it probably consists of about 30–45% water ice (including relatively small amounts of volatile ices), with the remainder being rocky material. Triton's surface area is 23 million km$^2$, which is 4.5% of Earth, or 15.5% of Earth's land area. Triton has a considerably and unusually high albedo, reflecting 60–95% of the sunlight that reaches it, and it has changed slightly since the first observations. By comparison, the Moon reflects only 11%. Triton's reddish colour is thought to be the result of methane ice, which is converted to tholins under exposure to ultraviolet radiation.

(Excerpt end)

Observation:

Just like with Neptune, distant sunlight with its ultraviolet radiation is incapable of driving some behaviors.

Io and its geysers is mentioned. In electric cosmology, those are triggered by electrical discharges, not by a mechanism of extreme heating like geysers on Earth. Triton does not have an internal heat source to boil water for a geyser eruption.

# 17.10 Dwarf Planet Pluto

In 1930, Pluto was considered a planet. In 2006, it was changed to a dwarf planet.

Excerpt from Wikipedia:

Pluto (minor planet designation: 134340 Pluto) is an icy dwarf planet in the Kuiper belt, a ring of bodies beyond the orbit of Neptune. It was the first and the largest Kuiper belt object to be discovered.

The plains on Pluto's surface are composed of more than 98 percent nitrogen ice, with traces of methane and carbon monoxide. Nitrogen and carbon monoxide are most abundant on the anti-Charon face of Pluto (around 180° longitude, where Tombaugh Regio's western lobe, Sputnik Planitia, is located), whereas methane is most abundant near 300° east. The mountains are made of water ice. Pluto's surface is quite varied, with large differences in both brightness and color. Pluto is one of the most contrastive bodies in the Solar System, with as much contrast as Saturn's moon Iapetus. The color varies from charcoal black, to dark orange and white. Pluto's color is more similar to that of Io with slightly more orange and significantly less red than Mars. Notable geographical features include Tombaugh Regio, or the "Heart" (a large bright area on the side opposite Charon), Cthulhu Macula, or the "Whale" (a large dark area on the trailing hemisphere), and the "Brass Knuckles" (a series of equatorial dark areas on the leading hemisphere).

Sputnik Planitia, the western lobe of the "Heart", is a 1,000 km-wide basin of frozen nitrogen and carbon monoxide ices, divided into polygonal cells, which are interpreted as convection cells that carry floating blocks of water ice crust and sublimation pits towards their margins; there are obvious signs of glacial flows both into and out of the basin. It has no craters that were visible to New Horizons, indicating that its surface is less than 10 million years old. Latest studies have shown that the surface has an age of $180000^{+90000}_{-40000}$ years. The New Horizons science team summarized initial findings as "Pluto displays a surprisingly wide variety of geological landforms, including those resulting from glaciological and surface–atmosphere interactions as well as impact, tectonic, possible cryovolcanic, and mass-wasting processes."

According to the measurements by New Horizons, the surface pressure is about 1 Pa (10 μbar), roughly one million to 100,000 times less than Earth's atmospheric pressure. It was initially thought that, as Pluto moves away from the Sun, its atmosphere should gradually freeze onto the surface; studies of New Horizons data and ground-based occultations show that Pluto's atmospheric density increases, and that it likely remains gaseous throughout Pluto's orbit. New Horizons observations showed that atmospheric escape of nitrogen to be 10,000 times less than expected.

(Excerpt end)

Observation:

Pluto and Charon are in synchronous orbit, so Pluto rotates once in 6.4 days while Charon takes 6.4 days to orbit around Pluto at a barycenter distance of 17,181 km.

In other words they rotate face to face taking 6.4 days for once around.

It is a surprise after only one fly-by, we can estimate the age of its surface and have even identified several "processes'" which resulted in this surprising surface, including "obvious glacial flows."

With Chiron in synchronous orbit there is no periodic tidal force on Pluto's surface.

On Earth, glaciers flow by the accumulation of snow at the top so they, by their weight, slowly slide to a lower elevation by gravity.
On Earth, heavy glaciers can have their sea-bound end dragging and grooving the sea bottom.
On Pluto, its glaciers are said to flow in and out of a basin.

The weak atmosphere is probably incapable of moving enough material to freeze onto glaciers so they get heavy enough to move by only gravity.

On the image of Pluto features several grooves are shown. The "obvious glacier flows" mechanism is probably needed to explain grooves in the basin.

Erosion is not the only mechanism for creating surface features on a solid body. Many craters are formed by an electric discharge on planets, moons, and asteroids.

Perhaps the grooves on Pluto are electrical scarring such as some canyons found on the terrestrial planets.

In our solar system where the gas giants are moving about, perhaps a more likely explanation is the smooth icy surface was deposited from a gas giant rather than the ice molecules moving around Pluto's surface are within a very weak atmosphere. Nitrogen is in the atmospheres of Saturn and Titan. That might have implications for the history of Pluto.

17.10.1 Pluto round craters

Pluto, like our Moon has round craters and some have a peak at the center of a flat floor with a consistent radius to the vertical walls around the circumference

There is one image in References with several craters. Unlike any craters on the Moon, these walls have vertical striations.

Another image has a crater cluster.

Some features have "informal names" in this coarse resolution image, including several round craters.

These features probably have an electrical cause, by an electrical discharge perpendicular to the surface, rather than from an explosive impact having a random trajectory.

## 17.11 Asteroids

### Linear Cratering on Tiny Asteroid

Image and story from Wikipedia topic:
2867 Šteins

This ia remarkable asteroid imaged by the Rosetta fly-by during the 7 minute encounter in 2008.

This 5 km asteroid has 23 named craters. There are about 9 round individuals in a line on the right side of the image.

This must have been an exciting sequence of electric discharges with the asteroid rotating during the encounter. This asteroid still rotates every 6 hours.

These craters require an awkward explanation for impacts.

### Cratering on Peaks of Asteroid

Image and story from Wikipedia topic titled:

21 Lutetia

21 Lutetia asteroid was imaged by the Rosetta fly-by during the encounter in 2010, getting 462 images of 50% of the surface (while at 15km/s).

Wikipedia has an image. This 100 km asteroid has craters and grooves.

The Rosetta mission(ESA) has a page titled:

Asteroid (21) Lutetia

This page includes annotated images.

The 2 most interesting round craters are:

1) just left of center (named Bonna in ESA image)
This is on the high point.
2) at the bottom (named Patavium in ESA image)
This is another high point.
Also interesting is the rim shot just above to the left of the bigger one. (named Bagacum in ESA image)
There was a round crater formed, and then another discharge hit on its rim.

There are many little, round, sputter craters from the encounter. This asteroid still rotates every 8 hours.

The description offers 7 geological regions about the terrain and impacts as well as the assumed history of this asteroid. Its dust layer might be 3km thick.

Lightning bolts prefer a high point.

## 17.12 Comets

A comet coma is an electrical discharge around an asteroid. It is not water vapor from ice.

Excerpt from Wikipedia:

A comet is an icy, small Solar System body that, when passing close to the Sun, warms and begins to release gases, a process called outgassing. This produces a visible atmosphere or coma, and sometimes also a tail. These phenomena are due to the effects of solar radiation and the solar wind acting upon the nucleus of the comet. Comet nuclei range from a few hundred meters to tens of kilometers across and are composed of loose collections of ice, dust, and small rocky particles. The coma may be up to 15 times Earth's diameter, while the tail may stretch beyond one astronomical unit. If sufficiently bright, a comet may be seen from Earth without the aid of a telescope and may subtend an arc of 30° (60 Moons) across the sky.

(Excerpt end)

Observation:

Probes have visited several comets and none matched that description. All are solid bodies with no ice.

Thunderbolts Project video was titled:
Electric Comets

It described recent probes to several comets. Every probe found surprises and little if anything conforming to expectations.

One interesting conclusion is the observed presence of water is from chemical reactions, not from a "pool" of water on a primordial body. In other words, the comet makes water.

Despite these findings, the theory of an icy snowball as a comet persists through no evidence for it.

# 19  Final Conclusion

This third book by the author offers further suggestions in the progression of cosmology beyond those steps described in the first two books.

The 3 books offer suggestions for improving modern cosmology, which made a number of mistakes before 1930 and never adopted the work of Hannes Alfven. He was awarded the 1970 Nobel Prize in Physics for plasma physics, which is important on Earth but also on the cosmological scale.

This $3^{rd}$ book provides more details about electromagnetic radiation and electric currents which generate magnetic fields.

One of the major problems in cosmology is the neglect of synchrotron radiation. Its frequency range is from radio to gamma ray. Cosmologists consistently offer unjustified temperatures to explain everything as thermal radiation, which is unfortunately limited to the range of infrared to ultraviolet.
Nearly every observation of an object or even a diffuse cloud, with X-ray, the explanation is consistently an impossible temperature. There is a failure to recognize thermal radiation spans between only ultraviolet and infrared. Rsdio sources are mentioned with no cause, missing another part of the synchrotron radiation range of frequencies.

This mistake follows the neglect of plasma behaviors and electromagnetic forces noted in the earlier books.

Together, the 3 books offer suggestions to change much of modern cosmology.

# 20 References

The references in this book are available as clickable links from a page in the author's web site.

1. Start web browser.
2. Go to this site: www.cosmologyview.com
3. Make sure the browser is on the correct home page: **Cosmology Views**

4. Scroll to near the middle.
5. Select: **Books by the author**

This page presents information for each book.
Locate the columns for this book.

6. Locate: **Cosmology Connections**

7. Below it, locate the date of this book's edition: 09/17/2020

8. Select: **References** after the correct date.

The selected page will list the references in the book by page number, with a link to that reference.

www.ingramcontent.com/pod-product-compliance
Lightning Source LLC
Chambersburg PA
CBHW071350210526
45465CB00001B/49